Understanding the Current Global Socio-economic and Ecological Crises

V. Ratna Reddy · Bhagirath Behera

Understanding the Current Global Socio-economic and Ecological Crises

In Search of an Alternative Integrated System

Springer

V. Ratna Reddy
Livelihoods and Natural Resources
Management Institute
Hyderabad, Telangana, India

Bhagirath Behera
Department of Humanities
and Social Sciences
Indian Institute of Technology Kharagpur
Kharagpur, West Bengal, India

ISBN 978-3-031-98599-7 ISBN 978-3-031-98600-0 (eBook)
https://doi.org/10.1007/978-3-031-98600-0

This Springer imprint is published by the registered company Springer Nature Switzerland AG
The registered company address is: Gewerbestrasse 11, 6330 Cham, Switzerland

Unregulated free markets actually reduce economic opportunities for the majority and help individuals to siphon wealth from the many to the few.

(Joseph Stiglitz, 2024).

Foreword

The contemporary world faces multifaceted and interrelated crises, including social disintegration, economic disparity, and environmental degradation. These systemic challenges, exacerbated by climate change, accelerated urbanization, technological disruption, and political volatility, have exposed the inadequacies of prevailing development paradigms that tend to privilege economic growth over equity, prioritize profits over people, and endorse resource exploitation at the expense of ecological sustainability.

This volume, *Understanding the Current Global Socio-economic and Ecological Crises: In Search of an Alternative Integrated System*, provides a foundation for societies to reorganize economic activity, govern natural resources, and advance human well-being. The current growth-centric and self-interest-driven economic model, reliant on the unsustainable extraction of natural capital, is fundamentally flawed. This book advances the case for a more balanced development paradigm that harmonizes economic advancement with social equity and environmental stewardship. Drawing on post-growth economic theories and diverse philosophical traditions, the volume investigates the assumptions of mainstream economic thought and explores viable alternatives.

While capitalism has been instrumental in generating material prosperity, it has disproportionately prioritized financial capital over human and ecological well-being, often externalizing social and environmental costs. Although technological innovation has been a catalyst for economic growth, it has also contributed to widening socio-economic inequalities and ecological degradation. These challenges often stem not from inherent technological flaws, but from exploitative and profit-maximizing deployment practices. Furthermore, environmental discourse is frequently politicized, particularly in North–South geopolitical dynamics, thereby hindering equitable and effective climate action.

The authors underline that nature-based solutions (NBS) provide a promising avenue for restoring balance between ecological and financial capital. By emphasizing the conservation and enhancement of ecosystem services, such as soil fertility, water purification, and biodiversity, NBS aligns with sustainable agricultural practices and broader ecological strategies. However, their widespread adoption is impeded by factors including long implementation timelines, undervaluation of non-market outputs, and limited institutional uptake. Unlocking the full potential of NBS will require integrated, interdisciplinary research and robust policy incentives that reflect both their ecological and economic contributions.

This manuscript points that the technological innovation and institutional reform alone are insufficient to address the depth of the current crisis. What is needed is a profound transformation in individual and collective behavior across systems of production, consumption, and governance. This includes a paradigmatic shift from consumption driven by greed to one informed by need, drawing inspiration from ancient wisdom and eastern philosophical traditions. Sustainable development thus necessitates a re-evaluation of societal values, individual choices, and ethical frameworks—not merely policy adjustments or market-based solutions.

This volume is intended as a resource for scholars, policymakers, development practitioners, and students who are committed to identifying durable responses to the complex crises facing humanity. Through critical engagement and interdisciplinary collaboration, the book aims to contribute meaningfully to the global pursuit of development pathways that are viable, just, inclusive, and ecologically sound. This is a must-read book for policymakers, researchers, and students.

Dr. Dil B. Rahut
Senior Research Fellow
Vice-Chair Research
Asian Development Bank Institute
Tokyo, Japan

Preface

This book is the outcome of our long-standing concerns about the global development process. To what extent have growth-centric approaches and free-market systems contributed to reducing global poverty and improving living standards? The effectiveness of the trickle-down model in enhancing well-being and equity across populations remains questionable, with increasingly ambiguous answers. The benefits of economic (GDP) growth have been limited and distributed unevenly, both across and within countries. More critically, current growth is driven by the unsustainable exploitation and degradation of natural capital. The continued deterioration of natural systems is already undermining both current and future growth prospects globally, with visible impacts now evident even in developed nations. The changing geopolitical landscape and evolving socioeconomic dynamics underscore the fact that the prevailing economic model is neither effective nor sustainable.

While this book offers a critical analysis of the growth-centric approach, it does not entirely dismiss it. We acknowledge that the current selfish motive driven economic philosophy has contributed to humanity's material prosperity. However, it has failed to ensure equitable outcomes and has compromised the natural capital needed by future generations. The foundation of profit-making has shifted from exploiting human capital to overexploiting natural resources. Drawing on ancient Indian scriptures, philosophical traditions, and cultural ethos, the book identifies fundamental flaws in modern economic thinking, which is often driven by self-interest, profit maximization, and exploitative, exclusionary practices. These include a disregard for biodiversity in production processes, unrestrained consumerism, greed, and entrenched biases in governance structures, including crony capitalism.

The book argues that only behavioural changes at a fundamental level can address the deep-rooted malaise in the present system. Such behavioural changes can be brought about through the right kind of education and the promotion of the right livelihoods at the societal or community level. Economic philosophy must shift away from a selfish motive paradigm, where the individual is dominated by self-regarding behaviour, primarily focused on accumulation, pleasure-seeking consumption, and greed, and move towards an orientation in which economic agents make critical daily decisions with a community, regional, and global outlook. This shift ensures that individual actions not only serve personal interests but also contribute to the welfare of others.

Although the book draws from ancient Indian scriptures—sometimes regarded as religious texts—the arguments are developed on the basis of people's way of life and daily livelihood and cultural practices during historical/ancient periods, not on religious dogma. In this sense, we are drawing lessons from historical evidence rather than religious preaching. Similar insights into socioeconomic well-being and compassionate governance can also be found in other ancient texts, particularly those written after the Christian era.

The ideas presented in this book have been shaped through years of rich discussions with colleagues, friends, and family. We are deeply grateful to Dil Bahadur Rahut for generously reviewing the manuscript and for writing the foreword. Special thanks goes to T. Chiranjeevi for her insightful reflections on ancient Indian texts, eastern philosophy, and psychology. Our conversations with Anurag Krishna Vippala about key post-growth literature were instrumental in framing our arguments.

Hyderabad, India V. Ratna Reddy
Kharagpur, India Bhagirath Behera

Competing Interests The authors have no competing interests to declare that are relevant to the content of this manuscript.

Contents

Chapter 1
Introduction: Growth Versus De-Growth

1 Poverty of Nations

Global economies are at a crossroads, and there are no exceptions among the countries. Even if some countries appear to be performing better at some point, it is only cyclical. Global economies have not prospered much despite the improved growth rates in gross domestic product (GDP) over the past five decades. This is very well reflected in the number of countries in the category of developed and developing countries. The post-world war division of countries as developed and developing or North/South has perpetuated with an add-on category of emerging economies. Ironically, very few countries (South Korea) have moved from the status of developing to developed over the past five decades. At the same time, countries have been downgraded from developed to developing or emerging (Soviet bloc countries), and some are at the margin (including some countries in Western Europe and Arab nations). A deeper socioeconomic assessment would reveal that more countries have slipped in their socioeconomic status than the number of countries that have improved their status. Across the world, economies (developed, developing and emerging) are on a regressive path. This is not only evident in the economic sphere but also in the context of sociocultural, political and ethical spheres. The decline is omnipresent irrespective of political ideologies and economic philosophies. While the countries that adopted socialist political ideologies (Soviet bloc) were the first to go down during the 1990s, capitalist (The USA and Western Europe), mixed (India) or transformed (China) ideologies have faced the downturn in recent years. Similarly, democracy, as well as other forms of governance, are not spared from the process. Because most countries across continents are going through socioeconomic crises in one form or another.

Traditionally, the socioeconomic development of a country is measured in terms of per capita annual income (GDP). Per capita income, in turn, is used to define the poverty line to measure the extent of poverty. The poverty line varies from US$

V. R. Reddy and B. Behera, *Understanding the Current Global Socio-economic and Ecological Crises*, https://doi.org/10.1007/978-3-031-98600-0_1

2.5/per capita/day (developing countries) to US$ 30/per capita/day (developed countries) across the countries depending on their food basket coupled with their relative socioeconomic development. Historical studies have shown that globally extreme poverty has declined from about 75% to about 10% over the past 200 years (Roser, 2021). This is mainly due to the adjustment of poverty lines, which are fixed at US$ 2.15/day across the countries. At the given poverty line of US$ 30/per capita/day most of the developed countries have a poverty rate of less than 1%. At the global level, the rate of extreme poverty is estimated at 84% when the poverty line is set at the developed countries level, i.e., US$ 30/per capita/day. This is more than that of the estimated poverty rate of about 75% in 1820 (Roser, 2021). In terms of the number of extremely poor, the increase over the past 200 years is much higher i.e., 75% of the 1 billion population in 1820 against 10% of the above 8 billion population. Considering the differences in prices and consumption baskets the poverty line for the developing countries could be lower than that of developed countries. Global extreme poverty is estimated at 58% with a lower poverty line of US$ 10 per capita/day (Roser, 2021). This measure of poverty is purely based on per capita income levels. Keeping price differences across the countries/regions purchasing power parity (PPP) has been used to measure relative poverty across countries since the 1990s.

Going by the law of averages the per capita income (GDP) measure of poverty is clearly biased. These estimates of poverty do not consider income inequalities within the countries. Though global inequalities declined in the middle of the twentieth century, they started increasing after the 1990s in most countries (Hasell, 2023). It is observed that inequalities are rising more sharply during the post-pandemic years (after 2021). Besides, income-based assessment is also not comprehensive enough to capture all dimensions of poverty. New approaches to poverty assessment have been introduced since the 1980s, viz., human development index (HDI), multi-dimensional poverty index (MPI), etc. These approaches focus or give more weightage to the non-income dimensions of poverty. While HDI considers income as an indicator, MPI does not include income explicitly. HDI has shown a progressive improvement in most countries since the 1990s (Bastian & Pablo, 2023). MPI has also recorded a substantial decline in the recent decades since its adoption (UNDP, 2023). HDI and MPI are now widely adopted across the countries though per capita income remains the core indicator of extreme poverty measurement in most countries. For, most of the other measures do not have the flexibility of fixing and changing the poverty line as per the country or regional characteristics.

While the rate of poverty is declining across the nations, the number of poor is swelling especially in developing countries, irrespective of the approach used. Recent developments clearly indicate that even developed countries are experiencing a rise in hunger and poverty, as food inflation has touched historical highs, especially after the pandemic and the Ukraine war. It is observed that even at US$ 30 per capita/day, developed countries have been experiencing an increase in the extent of poverty. Between the1980s and 2022, the extreme poverty rate has increased from 0.5 to 1.5% in the USA, from 0.1 to 1% in the UK, and from 0 to 0.4% in Denmark (Hickel & Sullivan, 2024). The experience of these countries illustrates that even high levels of growth and aggregate output with higher per capita incomes are not able to meet the decent living standards (DSL) of the people. In this context, it is argued that increasing the poverty line for developing countries may require higher growth rates, resulting in resource degradation.

Another important dimension that is missing in the poverty assessment is the quality of basic needs in developing countries. Not only the nutritional value of the food basket is declining rapidly but also getting increasingly contaminated with agrochemicals (Reddy & Rahut, 2024). Added to this is the decreasing quality of some basic needs like water, sanitation and hygiene (WASH), health and education services. This would further aggravate the burden of poverty, as the cost of basic needs keeps rising. Thus, extreme poverty of nations has deteriorated over the decades by any dimension or standard. Though hunger and destitution are less conspicuous in the present-day anatomy of poverty is more complex, as it considers multiple aspects that are beyond just food and hunger. That is, the malice is much deeper and clouded over by superficial articulations in terms of GDP and stock market indices.

2 Development and Degradation

The decline in the extent of poverty (percentage of poor) through high GDP growth, aggregate output and incomes has been achieved at the cost of natural resource extraction and environmental degradation. Globally, between 1992 and 2014, per capita produced capital increased by 100% and per capita human capital increased by 13% while per capita natural capital (forests, fisheries, minerals, etc.) declined by 40% (Fig. 1). These trends are likely to deepen with the increasing inequalities and unsustainable resource extraction. For, some of the resources are being extracted beyond sustainable limits causing irreversible damage to the natural system. The negative impacts of environmental degradation on the extent of poverty are evident in a number of countries, including the most developed countries like the USA. Historically development is associated with resource degradation, as these two are critically linked. The present developed world had gone through this during the post-World War era, when countries like the UK had experienced extensive degradation of their forest and water resources. Some of these countries could reinvest in rejuvenating their natural resource base along with a shift to alternative sources of energy (Scandinavian countries, Germany, the UK, etc.). Globalization of economies in recent decades has facilitated the shift of polluting industries to developing countries like India, China, Vietnam, etc. This results in a double burden of resource degradation for their own as well as for meeting the developed countries' output requirements. This has a multiplier effect on income levels, consumption, and basic needs demands that put pressure on natural systems. The increased demand for output and GDP growth is likely to severely damage the natural resource base in developing countries and adversely impact ecosystem services. For the developing countries are not capable of adopting the resource rejuvenation measures or shifting to alternative means. Developing countries need to clock sustained high GDP growth rates for a number of

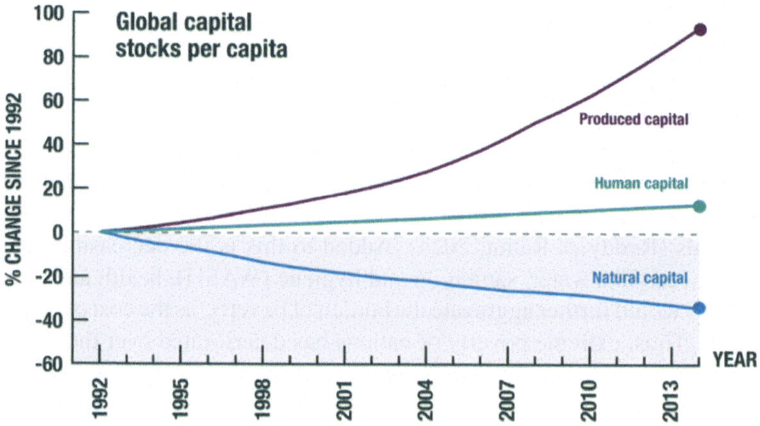

Fig. 1 Trends in produced, human and natural capitals (per capita). *Source* Managi and Kumar (2018)

years in order to make huge investments in reverting and strengthening the quality of natural capital. The fragile resource base in most of these countries will not be able to support and sustain such high GDP growth rates and output production.

Most developing countries are faced with increasing population (numbers) and declining agricultural productivity due to the degradation of natural resources like land, water, and forests (World Bank, 2021). With the added problem of climate risks, production stability, food security, and income are getting negatively affected. This is more conspicuous in the countries that adopted agriculture intensification, which has increased the pace of resource degradation. Crop intensification with green revolution technologies has not only degraded the resources but also impacted the nutritional value of food. Further, agrochemical contamination is on the rise, resulting in severe health impacts (human and livestock) in a number of countries. In fact, resource degradation is the most common and overriding factor across countries irrespective of demographics, resource base/drivers of growth, and modes of economy.

2.1 Demographics

The development or lack of it and the associated resource degradation are often attributed to the population pressure. Estimates clearly indicate that the present population with its present living standards (global average of US$ 20,000 per capita/year-2011 prices) can't be supported by the existing natural capital (Dasgupta et al., 2023). The existing natural system can support only about 3.3 billion population at the present consumption levels. Inversely, the natural system can support the existing population (say 9 billion) only at a consumption level of US$ 11,800 per capita/year

(2011 prices). Though these estimates may not be accurate enough due to methodological problems, they are indicative of the carrying capacity of the global ecological system. Population size is also linked to a lack of development and poverty across nations. That is, the larger the population, the lower the development, and the higher the poverty and resource degradation. It is argued that meeting the food demand for larger populations puts pressure on the natural resource base. But this argument does not hold good if one takes the per capita measure of consumption or emissions. Developed countries with lower population densities tend to consume and emit more even in absolute quantities. At the same time, none of the developing countries that house the majority of the poor are anywhere close to the threshold level of consumption (US$ 11,800) that is required to sustain the existing population.

The structure and composition of the population are also critical in influencing the trajectories of produced, human, and natural capital. In this regard, the proportion of the working and dependent population determines the potential of produced capital and dependence/degradation of natural capital. Countries like India have the advantage of having a greater proportion of the 15–60 years age group (working) population, which is termed as demographic dividend. European countries are in a disadvantageous position in this regard. Even China is slipping into the situation of rising the share of the dependent population. Countries with a higher proportion of the dependent population are characterized by declining growth rates coupled with the increasing burden of social security payouts. This is the main reason for the declining prosperity and quality of life in some developed countries. While most of the developing countries are favourably placed in terms of demographic dividend, they are not well placed as far as the quality of human capital is concerned. That is, skill sets, literacy, and education levels are at lower levels, affecting workforce productivity and efficiency adversely. Low productivity and poor efficiencies impact the resource allocation/use and governance practices, resulting in aggravating the degradation process. Thus, demographic structure and quality impacts the GDP/income levels and the associated resource requirement.

2.2 Growth Drivers

The economic transition from agriculture to industry and from industry to the service sector is well-established in the mainstream theories of economic development. Economic transformation is expected to progressively move from agriculture base to industry and to service sectors. This transformation is measured in terms of sectoral shares in GDP as well as the share in people (employment) moving from farm to non-farm sectors/activities and rural to urban areas. Though the process of economic transition is expected to move systematically from agriculture to industry to services, a number of developing countries are experiencing a spurt in the service sector ahead of industry or manufacturing. While the contribution of agriculture to GDP is going down at a much faster rate, the share of agriculture-dependent population is declining at a much slower pace. The process of structural transformation is very different for

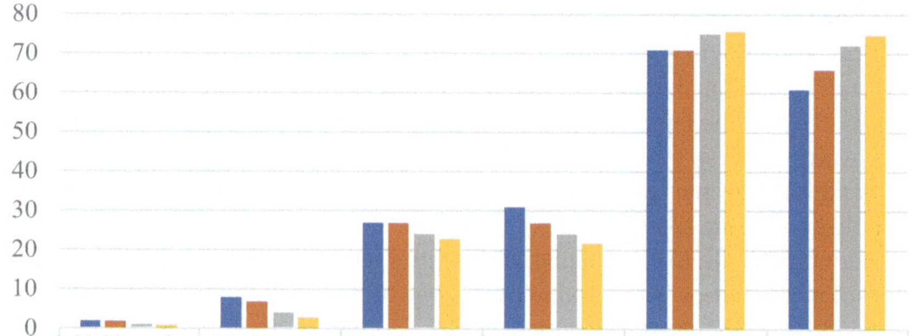

	Agriculture + allied		Industry		Service	
	% Share in GDP	% share in Employment	% Share in GDP	% share in Employment	% Share in GDP	% share in Employment
1990s	2	8	27	31	71	61
2000	2	7	27	27	71	66
2010	1	4	24	24	75	72
2022	1	3	23	22	76	75

Fig. 2 Structural transformation of high-income economies (1990s to 2022/3). *Source* World development indicators database for GDP ratios and for employment ratios: (https://data.worldbank.org/indicator/SL.AGR.EMPL.ZS?locations=8S)

middle and low-income countries when compared to high-income countries (Figs. 2 and 3). More importantly, even this slower decline is due to distress factors rather than preferred, i.e., push instead of pull factors. For instance, in Asia, only three countries (Japan, South Korea, and Taiwan) can boast of having experienced a smooth transition (as explained in development economics theories) from rural to urban or farm to non-farm sectors. In the rest of the countries, especially in South Asia, the distress is evident from the swelling of the number of farms and shrinking farm size over the years (Fan & Rue, 2020).

None of the developing countries have the capacity (in terms of investments and productive job creation) to support the rural–urban transition, given the magnitude (numbers) of the problem. Most of the employment generated for the rural poor (low skills) is in the service sector. The declining dependency on agriculture and increasing urbanization in most developing countries need not be seen as a positive development. The declining rate of extreme poverty (US$/capita/day) is often attributed to this distorted transition. One has to probe deeper (relative poverty, malnutrition, access to basic amenities like quality of water, sanitation, health and education) to understand the real malaise. This is clearly a sign of 'distress development'. Therefore, the potential for traditional development transition is very limited in most of the developing countries and the way it is happening is clearly not desirable.

Historically the transition from agriculture to industry and services has fuelled the GDP and output growth. Transition to the service sector is associated with higher development, as evident from Fig. 2. For the service sector, growth is also associated with higher labour productivity. At the same time, service sector-driven growth is

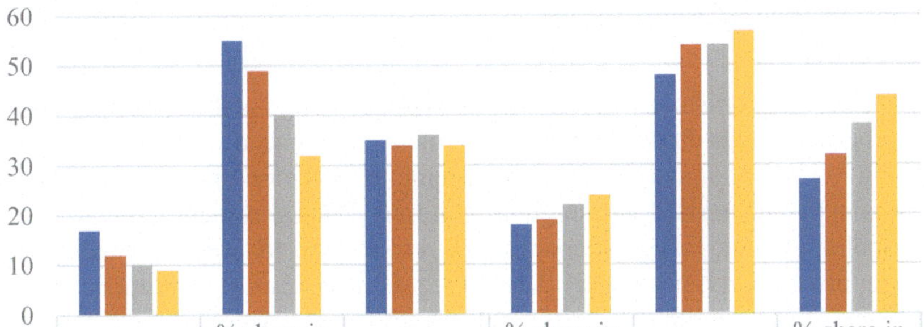

	Agriculture + allied		Industry		Service	
	% Share in GDP	% share in Employment	% Share in GDP	% share in Employment	% Share in GDP	% share in Employment
1990s	17	55	35	18	48	27
2000	12	49	34	19	54	32
2010	10	40	36	22	54	38
2022/23	9	32	34	24	57	44

Fig. 3 Structural transformation of middle and low-income economies (1990s to 2022/3). *Source* World development indicators database for GDP ratios and for employment ratios: (https://data.worldbank.org/indicator/SL.AGR.EMPL.ZS?locations=8S)

evident even in the middle- and low-income countries. For instance, countries like India, Vietnam, etc., are projected as success stories of service sector-driven growth. But, in the absence of an evolved manufacturing sector, the service sector is very slow in absorbing the labour force moving out of the agriculture sector. For, manufacturing is stagnant in terms of share in GDP and employment in the middle- and low-income countries. As a result, agriculture continues to accommodate a substantial proportion of the labour force in these countries though it contributes marginally to the GDP. The high growth rates achieved in countries may not be sustained in the long run unless the manufacturing sector is equally strengthened.

While agriculture and manufacturing sectors draw heavily on natural resources, services-driven growth is less dependent on the natural system. However, the services-driven high growth rates are observed to be unsustainable, as service sector growth is linked and dependent on growth in other sectors, especially manufacturing. It is observed in the case of South Korea that production-oriented services could only fuel sustainable GDP growth (Kim, 2006). Thus, services-driven growth does not support or sustain higher per capita incomes though it has a low ecological footprint. Conversely, a lower ecological footprint of services-led development means lower per capita incomes and compromising on the present standards of lifestyle.

2.3 Modes of Production

Feudalism, capitalism, socialism/communism, and a mix of them are the prime modes of production that modern civilizations have experienced over the past 200 years. While feudalism is associated with extreme exploitation and economic inequities, it was not characterised with high growth rates. This was mainly due to the limited technological innovations/adaptation for resource exploitation, enhancing land and labour productivity. It was only during the second industrial revolution (1850–1914) that the scale economies of new technologies were exploited, and the geographical concentration of technological leadership expanded to more countries, though limited to the Western world. During this period real wages and living standards have improved significantly. This has paved the path for many more industrial revolutions and capitalist modes of production (Mokyr, 1998). By the mid-twentieth century, capitalism and socialism/communism had replaced feudalism. Across the countries, capitalism is associated with democratic governance structures while socialism/communism is associated with dictatorial/fascist governance structures, though not many countries were under socialist regimes. Some countries like India have adopted mixed systems of democratic governance with socialist principles in their constitution.

In theory, capitalism is more exploitative (natural and human capital) and accumulative, while socialism is less exploitative and more distributive (equity). For, capitalism is based on neo-classical principles of self-interest and individual property rights. In contrast, socialist philosophy centred on community-based production and distribution. The practical experience of socialism does not vindicate any such theory of fair distribution, as the ruling class benefits the most from the toil of the proletariat. At the same time, capitalism is associated with high economic growth (GDP) and living standards though there are wide variations among the capitalist countries in this regard. Capitalism is based on the free market philosophy where markets, based on supply and demand for goods and services, determine prices and allocate resources. At the same time capitalism and free markets are driven by the neo-classical theory of self-interest behaviour. Though the self-interest-based free markets have been successful in fuelling growth to some extent, they failed in sustainable resource allocation (Fix, 2020). In fact, in their greed for higher growth and prosperity, the market systems have increased the pace of natural resource exploitation and degradation. More importantly, free market systems have failed to ensure or neglect the fair distribution of resources, growth, and prosperity. On the contrary, free markets and capitalist development have increased the inequalities over the years.

While most socialist systems, like the Soviet Union, have collapsed, countries like China have transitioned by adopting free market capitalism in specific sectors and regions by inviting foreign direct investment. Such a transition helped China to grow faster than the erstwhile market economies like the USA and Europe due to its authoritative governance structures. However, the high growth rates could not be sustained with the softening of international demand and also due to China's internal institutional and governance aspects. While countries like South Korea have done well by embracing capitalism in a democratic system, countries like India have adopted a middle path (public and private sector mix) and have not done that well despite the democratic setup. Overall, the free market systems have not only performed well but also sustained for a long when compared to other non-market systems like socialism/communism or regulated markets and public sector involvement. This is mainly due to the flexibility of the market systems that often adapt to the evolving socioeconomic conditions in the countries, while the other systems have been rigid in adapting to the changes.

Free market economies, however, are not conducive to the distribution of growth and prosperity across the socioeconomic sections of the population. Though the protagonists of growth believe in trickle-down impacts that would take care of equity, in reality, the trickle-down of growth has been slow and inadequate. Such inadequacies have become conspicuous even in developed countries like the USA, especially during and after the pandemic. This has led to the falling of effective demand due to rising inflationary pressures. Inflationary pressures are more severe in the food and basic needs sectors due to distorted production (supply) biases (Hickel & Sullivan, 2024) and also due to trade-related constraints on supplies. Handouts to the lower economic sections of the population are seen as an effective intervention for increasing effective demand. At the same time, strengthening the production systems (corporate sector) through the infusion of resources is also adopted to enhance supply. Though the handouts to poor households appear to be more effective than the infusion of resources to corporations, both are not sustainable in the long run. This is termed as fascist populism. In this regard, the Keynesian principle of increasing effective demand through employment generation (even if it means digging of wholes and filling them back) is likely to be sustainable rather than populism (hand-outs).

For, developing countries like India have been adopting the populist policies of free handouts to sustain the political regimes. The increasing political competition in a democratic set-up has led to ever increasing populism to unsustainable levels. In the process, political parties have been playing into the hands of corporates across the world over the years. This has led to the transition of capitalism into crony capitalism, which prioritizes the demands of the corporates rather than the basic needs of the common people (Aligica & Tarko, 2014). Crony capitalism compromises on the efficient allocation of resources through rent-seeking mechanisms and helps the accumulation of wealth in the hands of a few cronies. Though crony capitalism once was prevalent in developing countries of East Asia, south Asia, and Latin America, it has now spread to developed countries like the USA since the early 2000s with the bailing out of the banking and real estate sectors. Recent policies during the pandemic to support corporations in the name of economic stimulus emphasize the prevalence of crony capitalism. This crony capitalism may turn out to be the last straw of capitalist growth and sustenance. Besides, free markets in their yen for profits or profit maximization give the least priority to natural resource protection and their sustenance. This is no different in the non-market economies. That is, natural resources are at the core of creating wealth and prosperity.

While Marxism argued that the exploitation of human capital is the basis for profit maximization, the advent of technologies facilitated the exploitation of natural capital to fuel economic growth. That is, present-day growth is critically based on the exploitation of labour as well as natural resources. In fact, the extent of labour exploitation has increased many folds which is reflected in the corporate profits. Present-day entrepreneurs and their senior executives ensure their profits; they protect themselves from incurring any losses. For, no one has heard of a corporate CEO going bankrupt while corporates going bankrupt is very common. This, coupled with excessive and unsustainable exploitation of natural resources, is adding to the pace of surplus multiplication. As our preceding discussion clearly indicates, this is true across all contexts irrespective of market, resource base, and political systems. This is very much evident from the increasing climate risks across the world. These increasing climate risks are attributed to human disregard and mismanagement of natural systems across the countries. Degradation of natural resources has been a common factor in most countries that embarked on high growth paths. Though countries like Bhutan have chosen different paths, they may not be able to sustain themselves in the context of global onslaught on natural resource systems.

3 Looking Beyond GDP

The GDP-focused growth model has totally missed the development aspect. Income inequalities have been common across the countries; developing countries have additional problems of widening the gap between rural and urban areas/farm and non-farm sectors. For developing countries could not achieve sustained high GDP growth to productively employ the labour force released from agriculture in the non-farm sectors. Given the weak links between GDP growth and income distribution coupled with its stronger association with natural capital degradation, GDP growth could be a necessary but not sufficient condition for ensuring socioeconomic prosperity and well-being. This prompted a rethink about the GDP focused development and exploring the potential of alternative models or approaches in addressing the development dilemmas. Given the flexibility of the capitalist free market economies, a number of alternative approaches have been introduced and adopted at the policy level in order to address mainly the inequalities associated with GDP-focused growth.

Sustainable development (SD) is the first such alternative approach introduced and promoted at the global level under the United Nations umbrella. The seeds of sustainable development were sown in 1972 with the publication of 'Limits to Growth'. Sustainable development is a complex approach to integrating economic, social and environmental concerns that aims to address the needs of the present generation without compromising future needs. This requires long-term planning and implementation of different components of sustainable development. Hence, long term (over 20–30 years) sustainable development goals (SDGs) have been identified with time frames set for different countries. Given the long-term nature of SDGs, a number of quick fixes or short-term alternatives have been introduced at the policy level. Most of these could be considered as off-shoots of sustainable development or partially address some of the objectives of SD. Inclusive growth is one such alternative embraced by a number of developing countries like India.

Research has evolved well to integrate the multiple dimensions of sustainable development over the past three decades. Environmental and natural resource economics have become mainstream research, and a number of methods and tools have been developed and introduced to value and integrate natural resources into mainstream policy discourse. In the process, some important frameworks, viz., 'Water-Energy-Food Nexus'; 'Circular Economy', 'Life-cycle Assessment', 'Nature-Based Solutions', etc., which integrate long-term environmental externalities and impacts to arrive at final outcomes. At the same time, multi-disciplinary/ trans-disciplinary approaches came into vogue, which helped a better understanding of sustainable development. Though multi-disciplinary approaches are yet to be part of mainstream research, a number of organizations like the World Bank, FAO, OECD, ADB, etc., are adopting these approaches as part of their programs. Adoption of these approaches would certainly enhance our understanding of the problems associated with the GDP-focused approach; how far they could address the real problems of equity and natural capital depends on the integration of research understanding into policy. More importantly, as Bill Clinton, ex-president of the USA, once said "*It is the implementation-you stupid*".

Despite the comprehensiveness and integrated approach of sustainable development, the neo-classical philosophy of self-interest and GDP growth have remained central to sustainable development. Most of the SDG targets (poverty reduction, access to basic needs, etc.) are implicitly linked to growth and per capita incomes. Though sustainability is the main plank, there is no concrete approach or timeframe to strengthen the natural capital. The neoclassical approach of placing a value on natural resources is resorted to conserving natural capital. While value-based pricing could lead to efficient use of resources, it may not ensure resource sustainability as the returns from these finite resources continue to increase. Moreover, most developing countries are lagging behind in achieving SDG goals. The principles of sustainable natural resources governance, viz., reduce, rejuvenate/recharge, reuse and recycle are not receiving due priority at the policy level in most countries. For, most of the developing countries find it difficult to 'reduce' consumption or conserve resources, as these countries have yet to fulfil their basic needs like food and nutrition security. GDP growth and increasing per capita incomes continue to be the main indicators of global prosperity, despite the availability of alternative indicators like human development or multi-dimensional poverty.

4 De-Growth and Alternative Development Approaches

Over the past three decades, a number of alternative approaches to GDP and per capita income-focused development paradigms have emerged. Most of the alternative approaches are based on the de-growth school of thought. De-growth is a critique of GDP growth overshoot, appropriation, enclosure, and commodification that underpin the capitalist mode of production. On the other hand, de-growth promotes provisioning for human needs (use-value) through de-accumulation, de-enclosure and de-commodification. De-growth also puts greater value on labour and resources and rejects the racist ideologies that promote cheap labour and resources.[1] Some of the alternative approaches may be considered as the off-shoots of sustainable development. The majority of these approaches are focused on conservation and distribution. These include: rights-based approach, needs-based approach, flexible economy, solidarity economy, etc.

All these alternative approaches critique the neoclassical and GDP-focused approach that has self-interest as its core philosophy. Self-interest does not consider 'social good' or 'community welfare' while extracting from the common pool resources for the individual benefit. In the context of community-based development, self-interest often takes the form of free-riding, moral hazard, agency problems, etc., that vitiates group behaviour and strategies. At the same time, it is important not to ignore or consider 'self-interest' as a bad trait. In fact, 'self-interest' is an important driving force of global prosperity so far. It is argued that organized groups perform better when 'self-interest' is embedded. The problem with 'self-interest' is when individuals look beyond needs (not necessarily basic needs), and 'wants' become unlimited and turn into 'greed.' This calls for understanding and working on the behavioural aspects of individuals and groups so that production prioritizes and fulfils the needs of the majority rather than focusing on producing high-value goods for a few. The latter would improve the profit margin of the entrepreneurs, but impacts the effective demand adversely. Sustaining the effective demand of the majority population is the basis for economic well-being in the long run. In this context the alternative approaches become relevant.

Needs-based approach (NBA) looks at capitalist growth as distorted, as it doesn't address the needs of the real needy and deserving populations. NBA questions the nature and type of the present capitalist production as it is biased in favour of less-necessary goods and services. Instead, production forms that meet the basic human needs that would enhance human capital and capabilities need to be the priority. Such production forms should be fostered with improved and universal access to basic goods and services through public provisioning, if necessary. At the same time, countries, especially developed ones, should scale down the production of non-basic commodities. This would reduce the pressure on natural resource systems and bring resource use as per carrying capacity. The balanced production forms, as per the actual needs, could meet the demand without any substantial increase in total global throughput and output. It is estimated that decent living standards (DLS) for 8.5 billion people would require only 30% of current global resources and energy use, leaving a substantial surplus for additional consumption, public luxury, scientific advancement, and other social investments (Hickel & Sullivan, 2024). This could also be possible in developing countries as well through proper planning for the

[1] (https://www.sciencedirect.com/topics/social-sciences/degrowth#:~:text=Degrowth%20refers%20to%20a%20critique,excess%20throughput%20and%20colonial%20appropriation.).

adoption of efficient technologies, improved public services, and building sovereign industrial capacities.

Rights-based approach (RBA) is based on the UN human rights approach that endeavours the promotion and protection of human rights that include basic needs. The main focus of RBA is to address inequalities, discriminatory policies, and distribution of power that impede development progress. The absence of proper human rights in place is the main reason for sections of people getting left behind in the development process. Rights principles (universality, indivisibility, equality and non-discrimination, participation, accountability) are integrated into developmental policies, viz., right to development (civil, cultural, economic, political and social rights). RBA has become one of the guiding principles of the UN sustainable development cooperation framework.[2] Despite the UN's efforts to promote RBA across its member countries, RBA has yet to gain acceptance and find a place in country policies.

Deviating from other approaches *Solidarity Economy (SE)* integrates socio-political dimensions into development discourse. In the place of capitalistic self-interest and individual property rights, the SE framework advocates the practice of interdependence and collective discharge of economic activities producing basic goods and services. The underlying principles of SE include cooperation, democracy, social and racial justice, environmental sustainability, and mutualism. SE envisages the re-distribution of natural, economic and political resources to the hitherto deprived and vulnerable sections of the populations within a country and across the countries. While challenging the profit-oriented individual private systems, SE emphasizes the provision of real goods, viz., food, housing, education, health, etc. SE is expected to emerge from movements of social change integrating three common strategies: "*personal transformation, building alternative institutions, and challenging dominant institutions. Building solidarity economy movements requires building networks, federations and coalitions that align with SE principles and practices. This is where we become truly powerful*[3] *(p. 1).* SE is a more complex approach requiring fundamental changes like personnel transformation and building alternative institutions, which are possible through popular movements and very long-term in nature.

The flexible economy (FE) approach is based on the premise that the present trend of mindless economic growth is not only stretching the carrying capacity limits of the ecological systems but also the associated costs outweigh the benefits. FE argues that the existing de-growth approaches that emphasize conservation and shrinking of consumption levels are a difficult proposition for policymakers and the public, as they are not politically feasible. According to FE adaptability is the key to survival and progress. The economy should be adaptable to the changing ecological and social circumstances rather than remaining in a steady or fixed state. The main drivers of carrying capacity are consumption patterns and population size. These drivers can be regulated through behavioural changes viz., provision of scientific information and awareness building. FE calls for the involvement of ecologists, social scientists,

[2] (https://unsdg.un.org/2030-agenda/universal-values/human-rights-based-approach).

[3] (https://solidarityeconomyprinciples.org/what-do-we-mean-by-solidarity-economy/).

and economists in the economic planning process. None of the FE propositions are new, but it is argued that introducing flexibility and adaptability makes it marketable to policymakers and the public. The key to success i.e., 'saving ourselves from ourselves from the end-less growth' lies in selling the FE approach to public, business and political entrepreneurs.[4] It may be noted that adaptability and flexibility are also integral to capitalism. In fact, capitalism sustained itself for so long due to its flexibility and adaptability though the speed (geometric progression) at which it extracts and accumulates is much higher than its pace (arithmetic progression) to adopt.

All the above-discussed alternative growth models emphasize reducing the pace of present growth rates and keeping them within the sustainable limits of the carrying capacity of the ecosystem. While these alternative models look attractive, they fail to find acceptance at the policy level or attract public attention. Even popular approaches like 'Small is Beautiful' and 'Happiness Economics' could not make it to the mainstream policy discourse. While Schumacher's (1973) 'Small is Beautiful' suggests small-scale alternatives to development, Bhutan's 'Gross National Happiness' (GNP) during the 1970s was introduced as an alternative measure to GDP. Small is Beautiful is based on the Buddhist moral principle of '*attachment*' that is at the root of worldly problems. Detaching from the attachment to wealth, material goods, etc., would automatically curb the desire for accumulation leading to sustainable resource use. At the same time, it is not against wealth or material goods per se. It also draws on the Gandhian thought that nature has enough to support the needs of the people but not the greed of the people. The GNH is measured using four important indicators, viz., good governance, sustainable development, preservation and promotion of culture, and environmental conservation.[5]

Some of these approaches have the underpinnings of religious philosophies like Hinduism and Buddhism, though both these philosophies are often considered as a way of life rather than religions as such. Gandhian thought draws mainly from the Hindu philosophy. The Hindu philosophy or way of life is based on the '*Sanatana Dharma*' (eternal order) the longest surviving and practiced on earth. Being the oldest, it doesn't talk about any other approaches. In fact, Buddhism is a sub-set of the ancient Hindu way of life, which was adopted in a number of East Asian countries, including China. One of the reasons for this could be its flexibility and adaptability to the changing socioeconomic aspects over centuries. The Hindu way of life has provided equal importance to people, animals and nature. Even the feudal mode of production in India was considered less exploitative and more considerate of the working classes. In fact, Marx submitted that it is difficult to understand and explain the feudal relations in the Indian subcontinent. Conservation of resources and bio-diversity are ensured through the creation of use values for each and every species (flora and fauna). In fact, flora and fauna are revered as religious and cultural

---

[4] (https://medium.com/@martinrev21/the-flexible-economy-1ae4b6f08a07).

[5] (https://www.investopedia.com/terms/g/gnh.asp#:~:text=Gross%20national%20happiness%20(GNH)%20is%20a%20measure%20of%20economic%20and,of%20culture%2C%20and%20environmental%20conservation).

practices. *Sanatana Dharma* is not against self-interest, individual property rights and accumulation, though community or collective resource management practices have flourished for centuries. At the same time, greed is considered a vice and excessive use of natural resources is discouraged. Despite the cyclical changes in the governance structures over centuries, the cultural beliefs remain intact. At the same time, the onslaught of free market capitalism has eroded most of the cultural practices. In fact, India is fast moving towards crony capitalism and fascist populism (sourced from neo-liberalism) in recent decades.

Thus, all these alternative approaches to economic development are very appealing in the context of sustainable development. Protagonists of de-growth approaches consider curtailing growth is critical or the only option for achieving sustainable development across the countries. At the same time, how far these approaches can be translated into practical policy tools that help transform the system. The effectiveness of these approaches is critically dependent on behavioural changes at the individual, community and enterprise levels. Altering lifestyles and reducing consumption levels is not an attractive proposition for the public as well as policymakers. It is not politically feasible to implement policies that promote conservation and curtail present consumption levels. The trade-off between level/nature of production and food security/livelihoods and lifestyles needs to be understood and assessed. More importantly, none of these approaches have a strong theoretical basis or analytical framework to develop and design policies. Developing and designing an alternative approach with theoretical under pinnings and analytical support that minimises the trade-offs is a critical first step in this direction. This calls for an integrated approach incorporating economic, social, ecological and behavioural aspects. It is a difficult task but not impossible. If markets and economic instruments are capable of bringing in behavioural changes, so are the non-market and extra-economic factors or instruments.

The de-growth models are anti-growth for sure. Are these models also anti-development and anti-developing countries? While the developing countries are struggling to achieve the standards of the developed countries, the adoption of de-growth approaches would not only slow down the process but also result in these countries going backward and suffering in terms of food security and basic needs. It may not be fair to expect the countries that are suffering from poor availability of food and other basic amenities to reduce their consumption and lifestyles. The adverse impacts of high growth are expected to be borne by the developed world as well, but the likely impacts would be much less severe due to their advantageous position in terms of economic well-being. Moreover, it would be very difficult for the developed countries' populations to accept an almost 50% reduction in their lifestyles, i.e., a reduction from US$ 20,000 to about US$ 12,000 per capita/annum (Dasgupta et al., 2023). Thus, the de-growth options may not be a viable proposition under the present socioeconomic conditions across the world. This prompts some leading economists like Joseph Stiglitz (2024) to argue for and support the capitalist form of development, though in a different form. He terms it as 'progressive capitalism' in which markets as well as governments play important roles in the development process. He argues that free markets need to be regulated by the governments. Since markets can't correct themselves and provide social and environmental goods. Public investments, even in the area of technological development, are necessary for providing these goods and ensuring technological progression that benefits the majority of the population.

This book aims to provide an understanding of the potential and feasibility of the alternative approaches or development models that facilitate sustainable development. Specifically, the book attempts a critical review of important alternative approaches and philosophies and their acceptability at the policy and political levels. It assesses the trade-offs in the process of adopting de-growth models in place of earlier systems. And suggest an alternative approach that minimizes the trade-offs and makes it implementable at the policy level and palatable at the political level.

5 About the Book

This book is organized into five chapters. While this chapter provides the context and indicates the areas covered in the book. In the process, it discusses the rationale for the advent of post-growth approaches; the following chapters delve into the details of various post-growth approaches. Chapter Two reflects on the present economic systems and their repercussions on the wellbeing of the nations. It also critically discusses the role and limitations of technologies and institutions in achieving the overall economic development and well-being of the nations. Both industrial and agricultural, along with communication technologies are covered. In the case of institutions, formal as well as informal institutions dealing with economic and environmental aspects are dealt with. Chapter three provides a critique of some of the important post-growth approaches, viz., needs-based, decoupling growth and rights-based approaches. These approaches emphasise the behavioural changes towards altering consumption and production patterns. They focus on addressing the needs and rights of the vulnerable communities. The critique focuses on the practical aspects and the feasibility of adopting these approaches at the policy level. Chapter four deals with the modern and ancient approaches that focused on distribution and equity. It argues that modern systems like socialism failed not only in achieving the distributional objectives but also in sustaining themselves. At the same time, it provides some examples of solidarity, well-being and happiness economic models that are ecologically sensitive and economically distributive, though their scalability is a moot point. And these systems are yet to penetrate the broader policy discourse. On the contrary, ancient systems, based on Indian religious or ethical philosophies, could sustain for longer periods i.e., till they were disrupted by the so-called modern systems. These include *Hindu Dharma or Dharmanomics, Bhuddist*, and *Gandian* philosophies that emphasise on highly moral and ethical conduct and human behaviour in social and economic spheres of people's daily lives and economic activities. This chapter also identifies some of the root causes for the ills of the modern systems (self-centred, greed, etc.), i.e., the absence of the right education and right livelihood pursuits. And the final chapter draws on the ancient and recent Indian philosophies and explores the potential and possibilities of adopting an economic system that is beyond the premise of self-interest and greed.

References

Aligica, P. D., & Tarko, V. (2014). Crony capitalism: rent seeking, institutions and ideology. *Kyklos, 67*(2), 156–176.

Dasgupta, P., Dasgupta, A., & Barrett, S. (2023). Population, ecological footprint and the sustainable development goals. *Environmental and Resource Economics 84*, 659–675. https://doi.org/10.1007/s10640-021-00595-5

Fan, S., & Rue, C. (2020). The role of smallholder farms in a changing world. In S. Gomez y Paloma et al., (Eds.), *The role of smallholder farms in food and nutrition security*, https://doi.org/10.1007/978-3-030-42148-9_2

Fix, B. (2020). Economic development and the death of the free market, working papers on capital as power, No. 2020/01, forum on capital as power—toward a new cosmology of capitalism, s.l., http://bnarchives.yorku.ca/638/

Hasell, J. (2023). How has income inequality within countries evolved over the past century? Published online at OurWorldInData.org. Retrieved from: https://ourworldindata.org/how-has-income-inequality-within-countries-evolved-over-the-past-century

Herre, B., & Arriagada, P. (2023). The human development index and related indices: What they are and what we can learn from them. Published online at OurWorldInData.org. Retrieved from: https://ourworldindata.org/human-development-index

Hickel, J., & Sullivan, D. (2024). How much growth is required to achieve good lives for all? Insights from needs-based analysis. *World Development Perspectives, 35*, https://doi.org/10.1016/j.wdp.2024.100612

Kim, H.-J. (2006). The shift to the service economy: Causes and effects, institute for monetary and economic research. The Bank of Korea, K-Developedia (KDI School) Repository. https://core.ac.uk/reader/51180950

Managi, S., & Kumar, P. (2018). *Inclusive wealth report 2018: Measuring progress towards sustainability*. Routledge.

Mokyr, J. (1998). The second industrial revolution, 1870–1914. https://faculty.wcas.northwestern.edu/jmokyr/castronovo.pdf

Reddy, R. V. & Rahut, D. B. (2024). Future of small holder farms in South Asia. Report Submitted, Asian Development Bank Institute, Tokyo, Japan.

Roser, M. (2021). Extreme poverty: How far have we come, and how far do we still have to go? Published online at OurWorldInData.org. Retrieved from: https://ourworldindata.org/extreme-poverty-in-brief

Schumacher, E. F. (1973). *Small is beautiful: Economics as if people mattered*. Blond & Briggs (1973–2010), HarperCollins (2010–present).

Stiglitz Joseph, E. (2024). *Road to freedom: Economics and society*. Random House.

UNDP (United Nations Development Programme). (2023). 2023 Global multidimensional poverty index (MPI): Unstacking global poverty: Data for high impact action. https://hdr.undp.org/system/files/documents/hdp-document/2023mpireporten.pdf

World Bank. (2021). The natural resource degradation and vulnerability nexus: An evaluation of the World Bank's support for sustainable and inclusive natural resource management (2009–19). Independent Evaluation Group. World Bank. https://ieg.worldbankgroup.org/sites/default/files/Data/Evaluation/files/NaturalResourceDegradationVulnerabilityNexus.pdf

Chapter 2
Current Economic System and the Real-World Problems: A Reflection

1 Context

There is a stark disconnect between the current economic system and real-world issues. As the global economy expands, so do its associated problems. The prevailing economic model, centered on endless material growth, consumption, and profit maximization, threatens the ecological balance of our planet (Jackson, 2009; Sachs, 2015). Thus, balancing economic growth with environmental sustainability is one of humanity's most complex challenges (Basheer et al., 2022). Rapid population growth and increasing material consumption have led to severe environmental consequences, including anthropogenic climate change (Mukherjee et al., 2022), biodiversity loss (Cafaro et al., 2022; Crist, 2022), ocean acidification (Osborne et al., 2022), land degradation, threats to food security, extreme poverty, income inequality, and health hazards. The central question remains: How can humanity reconcile its expanding economic activities with the finite capacity of the planet's ecosystems? This chapter provides a detailed diagnosis of the present economic system.

For decades, scientists, economists, and scholars have explored sustainability. Their recommendations largely focus on two factors: (1) technological innovations and (2) institutional frameworks that promote the efficient use and management of natural resources. However, despite significant scientific and technological advancements, as well as policy and regulatory measures, economic and environmental challenges have continued to escalate. This underscores the need to reassess these issues, identify their root causes, and seek long-term solutions rather than merely addressing symptoms.

Governments worldwide prioritize economic growth as a key measure of prosperity. Economists and policymakers view it as essential for improving living standards and alleviating poverty (Perman et al., 2003). Gross Domestic Product (GDP)—the total market value of a country's output—remains the primary indicator of economic growth. GDP per capita, which represents economic output per person, is

V. R. Reddy and B. Behera, *Understanding the Current Global Socio-economic and Ecological Crises*, https://doi.org/10.1007/978-3-031-98600-0_2

commonly used to gauge living standards. However, a rising GDP does not necessarily equate to broad-based well-being, as it fails to account for income distribution and economic equity.

Economic growth, often equated with rising GDP per capita, does not necessarily translate into "development." True economic development requires improvements across various social and economic indicators, such as reduced income inequality, lower mortality rates, and higher literacy rates (UNDP, 1990). However, both growth and development frequently neglect environmental degradation, raising critical questions: How long can growth persist? How can economic expansion align with the planet's ecological limitations?

These questions have fuelled one of economic history's most significant intellectual debates—the relationship between economic systems and the environment. Thomas Malthus first addressed this issue in his 1798 work, An Essay on the Principle of Population, arguing that exponential population growth would outstrip food supply, leading to scarcity, famine, and conflict. David Ricardo, another classical economist, introduced the concept of diminishing marginal returns, predicting that population growth would force agricultural expansion onto lower-quality land, exacerbating food scarcity (Hussen, 2004). While Malthus emphasized absolute scarcity, Ricardo focused on relative scarcity.

The field of environmental and natural resource economics emerged in response to Limits to Growth (1972), which warned that unchecked economic expansion would lead to resource depletion, rising pollution, and eventual economic collapse (Meadows et al., 1972). The study highlighted that environmental constraints impose strict limits on development, necessitating a fundamental reassessment of the trade-off between economic growth and long-term sustainability.

2 Present Economic Paradigm and Its Outcomes

The core principle of the present neoliberal economic paradigm is the close relationship between consumption and economic growth (Fig. 1). It is clear that without increasing consumption of material goods, higher economic growth cannot be achieved (Jackson, 2009). In other words, the satisfaction of sensory pleasures and desires of consumers by consumption of material goods drives the growth of the production of economic goods. Producers provide more choices regarding the variety of goods and services for consumers to choose which may help satisfy their desires (Fig. 1). In this context, an important distinction needs to be made between human "needs" and "wants/desires" (Behera, 2012; Jena & Behera, 2017). Conventional economics stresses that the satisfaction of unending human desires is the primary function of an economy (Gough, 1994). Higher economic growth induced by incessant mass consumption employs the working population whose livelihoods are critically linked to producing economic goods and their trades in domestic and international markets. The dictums of the present economic system are that once the economy grows, it must not de-grow; once it trades, it cannot stop the trade. This

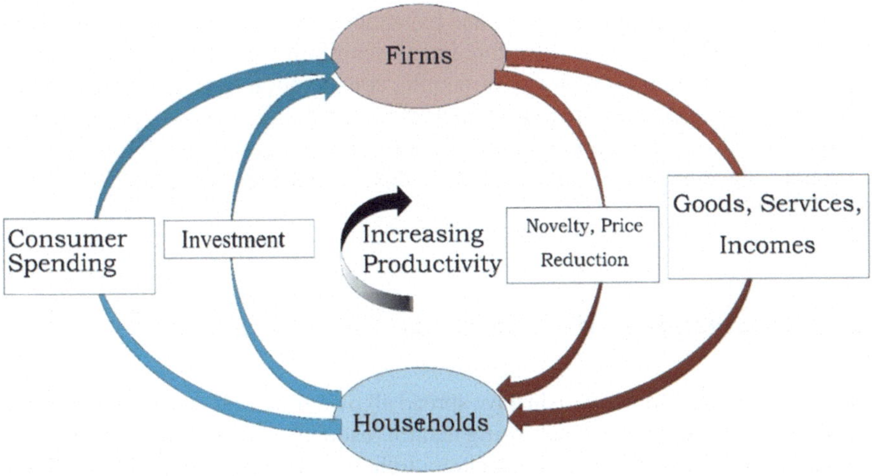

Fig. 1 The engine of economic growth: consumer-producer nexus. *Source* Jackson (2009)

paradigm actively alienates countries that do not pursue this paradigm and convinces people that there is no alternative to economic growth.

Hence, the modern economic system faces a profound dilemma: on the one hand, economic growth is not ecologically sustainable because the size of the global economy is large compared to the carrying capacity of the earth's ecosystem, on the other hand, as mentioned above, in the absence of positive economic growth large scale unemployment and other serious social consequences may result. Thus, the question that needs to be asked is: is infinite positive economic growth possible on a planet with limited resources?

The vicious circle of consumption and production of economic goods has led to a serious problem of environmental degradation and has threatened the sustainable well-being of both society and the earth's ecosystem (Behera, 2012, 2017). In other words, humanity has posed a serious threat to its current survival and future well-being. Since the Industrial Revolution, the world has registered a more than 200-fold increase in global output, which has caused immense damage to the global environment. Large-scale economic activity is rapidly and irreversibly changing the earth's climate, water cycle, nitrogen cycle, ocean chemistry, etc. It is observed that humanity is using so much land to accommodate growing economic activities that it is crowding other species of the planet, driving them to extinction (Sachs, 2015).

2.1 Ailments of the Current Economic Paradigm

The incessant consumption of goods and economic growth are dual causalities that threaten human well-being and natural ecosystems. Consumption of goods is promoted to secure people's physical and psychological well-being. Hence, individuals' fundamental freedom and welfare have been relegated to "only" consumptive choices. That is why for a majority of people, consumption is more important than sufficiency; production is more important than sustainability; growth of production of economic goods is more important than their distribution; accumulation/possession of economic goods is more important than their actual uses (Behera, 2012, 2013, 2017, 2023).

2.2 *Consumption is More Important Than Sufficiency*

Consumerism and commercialism have spread all over the world and have fuelled the growth of the production of goods using natural resources. To lead a healthy life, one requires basic needs such as food, shelter, clothes, and basic amenities. By having more disposable income, people tend to consume more than what is sufficient for their living. As we will elaborate on in the next section, it is not only the disposable income that encourages people to consume more than they need, but the urges/desires in the human psyche compel people to consume incessantly (Behera, 2017).

2.3 *Production is More Important Than Sustainability*

In order to meet the growing demand for goods emerging from the consumptive choices of people, large-scale production using sophisticated technologies is being undertaken. In addition to human resources production of goods requires inputs from nature in the form of biotic (renewable) and abiotic (non-renewable) resources. Renewable resources can become non-renewable if the rate of extraction is more than its growth rate (Perman et al., 2003). Large-scale production of goods using these natural resources can seriously undermine the ecosystem's sustainability, given the natural environment's limited capacity to regenerate itself. Large-scale production of goods is linked to consumer demand and economic profitability. Producers tend to seek and accumulate more profits, often at the expense of the natural environment.

2.4 *Economic Growth is More Important Than Distribution*

Modern economists have a very strong attachment to economic growth as a major policy objective because they see it as the only feasible way to solve the problem of poverty (Perman et al., 2003). The other route to poverty alleviation is through redistribution from the better-off to the poor, which is less preferred because generally, the better-off will resist attempts to redistribute from them to the poor, leading to social tension and violent conflicts. This path of higher economic growth has produced inequality in society; the top 20% of people own 80% of resources (Sachs, 2015). Most emerging economies are experiencing jobless growth, and the growth outcomes are not "inclusive" (ibid). Again, the primary reason for this outcome is rooted in the human psyche, which does not like to share resources with fellow human beings as self-interest dominates social interests.

2.5 Accumulation/Possession is More Important Than Use

The final ailment of the present economic system is the inherent tendency of people to accumulate/possess goods over and above their needs. This human tendency to accumulate more goods is to be safe and secure socially, economically, and psychologically. The tendency to acquire more includes both physical goods (more material goods) as well as psychological security (more social status, prestige, power, position, name, and fame). People tend to identify themselves with physical goods. Consumers always seek new products in order to be happy and secure their social status (Jackson, 2009).

The above discussion points out that both consumers and producers are "locked-in" viciously in a cycle that is becoming increasingly difficult to break. Consumers seek a higher level of satisfaction by wanting more economic goods and services. This incessant desire to satisfy can be both physical and psychological. It is dynamic in nature, as it is largely born out of comparison with other consumers, and comparison has no end. Producers whose sole aim is to maximize profits would like to capture ever-seeking consumers by providing them with a variety of goods at lower prices (as shown in Fig. 1). The challenge is to break this cycle and put the economy on a sustainable path.

Hence, in this regard, the key questions to be asked are: how can humanity's growing economic activities be reconciled with the limited capacity of the planet's ecosystem? Will technology outpace environmental constraints and prevent the collapse of the human population and economy? Will humans change their behaviour and thereby reduce pressure on the planet or rely on technology to meet their unending demands? What prevents people from exhibiting pro-social/pro-environmental behaviour? Humanities so far have attempted to break the above-mentioned vicious cycle and address the above questions by applying two approaches: Technological innovation and Institutional design. A brief discussion of these two approaches is provided below:

The above discussion points out that both consumers and producers are "locked-in" viciously in a cycle that is becoming increasingly difficult to break. Consumers seek a higher level of satisfaction by wanting more economic goods and services. This incessant desire to satisfy can be both physical and psychological. It is dynamic in nature, as it is largely born out of comparison with other consumers, and comparison has no end. Producers whose sole aim is to maximize profits would like to capture ever-seeking consumers by providing them with a variety of goods at lower prices (as shown in Fig. 1). The challenge is to break this cycle and put the economy on a sustainable path.

Hence, in this regard, the key questions to be asked are: how can humanity's growing economic activities be reconciled with the limited capacity of the planet's ecosystem? Will technology outpace environmental constraints and prevent the collapse of the human population and economy? Will humans change their behaviour and thereby reduce pressure on the planet or rely on technology to meet their unending demands? What prevents people from exhibiting pro-social/pro-environmental behaviour? Humanities so far have attempted to break the above-mentioned vicious cycle and address the above questions by applying two approaches: Technological innovation and Institutional design. A brief discussion of these two approaches is provided below:

3 Technological Innovations

Most people strongly believe that all the problems, including environmental degradation and poverty alleviation, can be addressed through innovation and better technologies (Tabash et al., 2025; Zhang et al., 2025). Technological development has helped tremendously to improve people's living standards by ensuring food security and other related basic necessities such as medical facilities, transportation, etc. (Pang, 2024). Modern technology has also helped humans extract ecosystem services and other critical facilities to maintain the current civilization.

These include improved communication, advancements in healthcare, enhanced access to education, economic growth, improved quality of life, sustainable environmental management, transportation, scientific research, workplace transformation, access to information, etc. Improvements in communication through mobile phones, email, and social media have made it possible for people to connect instantly, regardless of geographical location; video conferencing tools like Zoom and Teams enable remote meetings, fostering global collaboration and reducing the need for physical travel. Advancements in healthcare through technological innovation have led to the development of advanced medical devices, diagnostic tools, and treatments, improving patient outcomes and extending life expectancy; remote consultations and health monitoring have become possible, making healthcare more accessible, especially in rural or remote areas, which is popularly known as telemedicine. Enhanced access to education has been made possible through E-learning platforms which provide access to education for people worldwide, breaking down barriers of cost and location; technology has introduced interactive and personalized learning experiences, making education more engaging and effective.

Online platforms like Amazon and Alibaba have revolutionized the retail industry, providing convenience to consumers and opportunities for small businesses to reach global markets; automation in industries has increased productivity (Sharma, 2024), reduced costs, and led to the development of new job roles requiring higher skill levels, which have helped achieve higher economic growth and prosperity (Hall &

Helmers, 2024; Zhang, 2024). Improved quality of life has been achieved through smart devices like thermostats, security systems, and virtual assistants enhance convenience and security in daily life; streaming services, gaming platforms, and social media offer diverse forms of entertainment, making it easier to access content from around the world.

Sustainable environmental management through advancements in solar, wind, and other renewable energy technologies contribute to reducing carbon footprints and combatting climate change and sustainable practices in terms of technology enable more efficient resource management, waste reduction, and the development of sustainable products and practices (Qamruzzaman & Kor, 2024). Modern transportation, such as the development of electric and autonomous vehicles, is reducing dependence on fossil fuels and improving road safety, and GPS and traffic management systems help reduce congestion, save time, and reduce pollution (Ganguly et al., 2025). Big data and artificial intelligence (AI) are accelerating scientific research by analysing vast amounts of data quickly and accurately and advanced technology has enabled the exploration of outer space, leading to discoveries about our universe and potential future resources (de Zwart & Stephens, 2019; Rausser et al., 2023).

Technology has made remote work possible, providing flexibility and work-life balance for employees while reducing operational costs for businesses and platforms like Microsoft Teams facilitate teamwork and project management, improving efficiency and collaboration (Alegre & Pasamar, 2018). The internet provides access to vast amounts of information) enabling people to learn, stay informed, and make better decisions and AI-driven tools and algorithms help people find relevant information quickly and personalize their experiences online (Maclaurin, 1950).

In addition to services and its allied sectors, technological innovations have had profound impacts on both the industrial and agricultural sectors, driving productivity, efficiency, and sustainability (Achilladelis et al., 1990). These advancements have shaped the modern economy, addressing challenges and creating new opportunities in production, management, and resource utilization. A brief discussion on technological innovations in both sectors is provided below.

3.1 Technological Innovations in the Industrial Sector

The industrial sector has undergone several waves of innovation, from the first Industrial Revolution to the current era of Industry 4.0 (Morrar et al., 2017). These innovations have transformed manufacturing, logistics, and the overall supply chain. For instance, automation in manufacturing has reduced the need for manual labour and increased precision and efficiency (Aydalot & Keeble, 2018). Technologies like Computer Numerical Control (CNC) machines allow for precise cutting and shaping of materials with minimal human intervention. The use of robots in assembly lines has revolutionized the production process. Robots are capable of performing repetitive tasks with high accuracy, reducing errors and increasing productivity. Collaborative robots are designed to work alongside humans, enhancing productivity without completely replacing human labour. The integration of digital technologies into manufacturing creates "smart factories," which is termed Industry 4.0. These factories use IoT, artificial intelligence (AI), and big data to optimize production processes, predict maintenance needs, and reduce downtime. IoT devices enable real-time monitoring and control of machinery and processes. Sensors collect data from equipment, which is then analysed to optimize performance, improve safety, and reduce energy consumption. 3D Printing technology allows for the creation of complex components directly from digital designs. It is used in prototyping, small-scale production, and custom manufacturing. 3D printing reduces waste, speeds up the design process, and allows for more customization in manufacturing (Popkova et al., 2019).

Innovations in material science have led to the development of stronger, lighter, and more durable materials. These materials are used in various industries, from aerospace to automotive, to improve performance and efficiency. There is a growing focus on developing sustainable and environmentally friendly materials. Innovations include biodegradable plastics, recycled materials, and alternative composites that reduce the environmental impact of manufacturing (Harahap et al., 2023). AI is used to optimize production processes, from predictive maintenance to quality control. Machine learning algorithms analyse data from production lines to identify inefficiencies and suggest improvements. AI is also used to optimize supply chains, predict demand, manage inventory, and improve logistics. This reduces costs, increases efficiency, and enhances customer satisfaction (Harahap et al., 2023; Popkova et al., 2019).

3.2 Technological Innovations in the Agricultural Sector

The agricultural sector has also benefited significantly from technological advancements, often referred to as the "Green Revolution" in its earlier stages and now as "Agri-tech" or "Precision Agriculture" (Méndez-Zambrano et al., 2023). These innovations have increased crop yields, improved resource efficiency, and made farming more sustainable (Behera & Mishra, 2007). Global Positioning Systems (GPS) and Geographic Information Systems (GIS) enable farmers to map fields and monitor crop conditions with precision (Méndez-Zambrano et al., 2023). These technologies allow for site-specific management of crops, optimizing inputs like water, fertilizers, and pesticides. Drones equipped with sensors and cameras can monitor crop health, assess damage, and help in the precision spraying of pesticides and fertilizers. Satellite imagery provides large-scale monitoring of agricultural areas, helping in planning and disaster management (Ayim et al., 2022).

Biotechnology has led to the development of crops resistant to pests, diseases, and environmental stresses. GMOs can also be engineered to improve nutritional content and yield (Aksoy, 2016). Sensors placed in fields monitor soil moisture, temperature, and nutrient levels in real-time. This data is used to optimize irrigation, fertilization, and pest control, reducing waste and improving crop yields. Automated machinery, such as robotic harvesters, planters, and weeders, reduce labour costs and increase efficiency. Automated irrigation systems, controlled by sensors, ensure optimal water use.

Vertical farming involves growing crops in stacked layers, often in controlled indoor environments (Kalantari et al., 2018). This method reduces land use, water consumption, and pesticide use while increasing crop yields. Hydroponics and aquaponics are soilless farming methods that allow for the cultivation of crops in nutrient-rich water, reducing the need for land and water (Pantanella et al., 2010; Sayara et al., 2016). Aquaponics combines fish farming with plant cultivation, creating a closed-loop, efficient and sustainable system. AI and big data are used to predict weather patterns, crop diseases, and pest infestations, allowing farmers to take preventive measures. Predictive analytics also help in optimizing planting and harvesting schedules. Farm Management Software tools help farmers manage their operations, track expenses, monitor crop performance, and make data-driven decisions. This improves efficiency and profitability (Fountas et al., 2015).

Both sectors have seen significant improvements in productivity due to technological innovations. However, the industrial sector has been more rapidly transformed by automation and AI, while agriculture has seen more gradual changes, particularly in regions with less access to technology. Innovations in both sectors aim to improve sustainability. In industry, this is achieved through energy-efficient manufacturing processes, recycling, and the development of sustainable materials. In agriculture, sustainable practices focus on reducing resource use, minimizing environmental impact, and enhancing biodiversity. Both sectors face challenges related to adopting new technologies, including high initial costs, the need for skilled labour, and potential disruptions to traditional practices. In agriculture, there is also the challenge of ensuring smallholder farmers access these technologies.

Modern technology continues to evolve, and its impact on human life is profound, enhancing convenience, efficiency, and quality of life while addressing some of the world's most pressing challenges. However, technology has not been successful in effectively reducing environmental degradation. For example, the Jevons Paradox, named after the British economist William Stanley Jevons, is the counterintuitive observation that increases in the efficiency of resource use can lead to an overall increase in the consumption of that resource rather than a decrease. Jevons first noted this phenomenon in his 1865 book, "The Coal Question," where he observed that improvements in the efficiency of steam engines led to an increase in coal consumption in Britain.

Here's how the Jevon Paradox works: (i) *Efficiency Improvement*: A technological advancement or process improvement makes using a resource more efficient, meaning less of the resource is needed to produce the same output. (ii) *Reduced Costs*: The efficiency improvement typically reduces the cost of using the resource, both for producers and consumers. (iii) *Increased Demand*: The lower cost encourages increased resource use because it becomes more economically attractive (York & McGee, 2016). This can occur in two ways: (a) *Direct Rebound Effect*: The specific use of the resource becomes cheaper, so people use more of it. (b) *Indirect Rebound Effect*: The cost savings from the efficiency improvement are spent on other goods and services, which also require energy or resources to produce. (iv) *Overall Consumption*

Increase: The increase in demand can offset the savings from improved efficiency, leading to an overall increase in resource consumption (Giampietro & Mayumi, 2018).

Jevons Paradox highlights a potential challenge in addressing environmental and sustainability issues, as simply improving efficiency may not be sufficient to reduce overall resource use and environmental impact. Instead, it may be necessary to complement efficiency improvements with policies that limit resource consumption or promote alternative, more sustainable behaviours (Polimeni et al., 2015).

Using a graph (Fig. 2), Burger (2018) provided a conceptual model revealing how growth in the human system (population and economic growth) has been sustained and/or expanded by the positive feedback received from accumulated cumulative cultural evolution in terms of human experiences, technological innovation, and the increased flow of ecosystem services over the past several centuries. The author has also raised the same question, as it is raised above, as to whether this positive feedback can continue indefinitely on a finite planet. As aptly highlighted by Jevon Paradox, technological innovations may only help in extracting more ecosystem services from the planet; it may not help in removing scarcity of resources in an absolute sense and addressing the growing degradation of the natural environment.

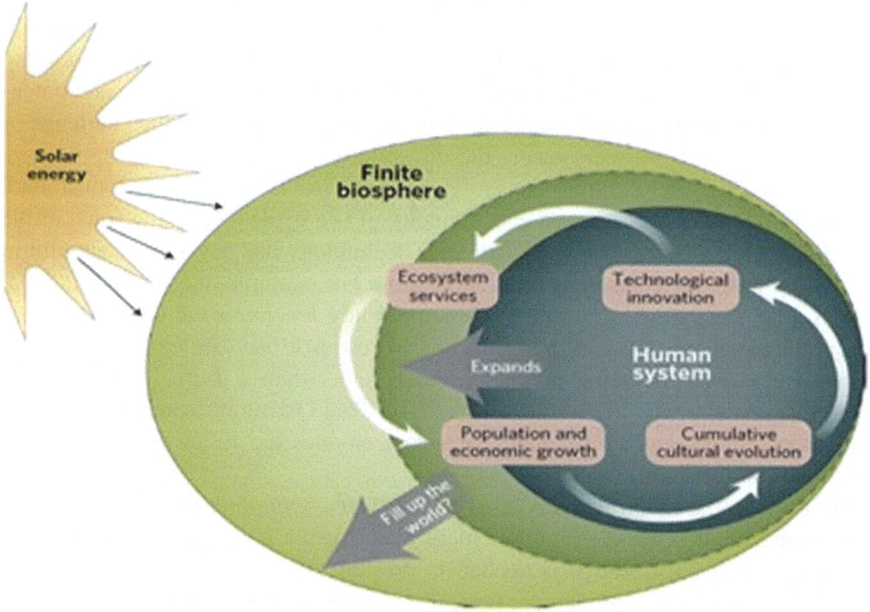

Fig. 2 Interaction between the human system (civilization) and the finite biosphere of the planet. *Source* Burger (2018)

4 Institutional Evolution and Design

Institutions are bridges between policies and implementation. Evolution and designing of appropriate institutions (e.g. laws and regulations) is another pathway that human society has been employing regularly and is found to be crucial in addressing both economic and environmental problems (Ebner, 2008). Ever since humans started living in settlements and forming society, humans have been developing various rules and regulations from time to time to better organize their social, economic, political, and moral/ethical behaviour for a peaceful, prosperous, and harmonious society (Neyapti, 2013). How these institutions have contributed to finding lasting solutions for both economic and environmental crises and their apparent failures in resolving growing economic and ecological problems are broadly discussed below:

4.1 Economic Front

The human system is currently being operated largely based on market mechanisms, a form of institution that human societies throughout the world have developed and been employing to organize their economic activities (Ebner, 2008). Markets are the backbone of economic activity, facilitating the exchange of goods, services, and resources (Loasby, 2001). They operate based on the principles of supply and demand, competition, and market efficiency, and are influenced by both internal dynamics and external factors like government policies and global events. Efficient markets allocate resources in the most productive way, ensuring that goods and services are produced and distributed at the lowest possible cost. The concept of market efficiency is central to economic theories and is often used to justify minimal government intervention. Understanding how markets function is essential for making informed economic decisions, whether as a consumer, producer, or policymaker.

A market is a platform or system where buyers and sellers interact to exchange goods, services, or assets. Markets can take various forms, ranging from physical locations like local farmers' markets to virtual spaces like online marketplaces. They play a fundamental role in the economy, determining the prices of goods and services through the forces of supply and demand. However, market mechanisms do not function in a vacuum (Wegner, 2003). It requires elaborate rules and regulations, and hence, governments often design rules and regulations in order to make the market function efficiently. For instance, laws and regulations are made in order to prevent monopolies, ensure fair competition, and protect consumers from exploitation. Governments maintain a stable and fair economic environment by enforcing antitrust laws and regulating markets. Similarly, in the financial sector, the government tries to bring financial stability by establishing regulatory institutions like central banks and financial watchdogs that enforce rules to ensure the stability of financial systems, preventing crises such as bank runs and economic recessions. Employment laws protect workers' rights, ensure fair wages, and prevent exploitative practices, contributing to a more equitable economy. International trade regulations help manage economic relations between countries, preventing trade wars and ensuring mutual benefits through agreements like tariffs, quotas, and trade partnerships (North, 2003).

As discussed above, while markets are typically self-regulating, government intervention is sometimes necessary to correct market failures, protect consumers, and promote fair competition. Governments levy taxes and provide subsidies in order to influence the behaviour of consumers and producers, such as taxing tobacco to reduce smoking or subsidizing renewable energy to promote its use. The government may provide certain goods and services that markets fail to deliver efficiently, such as national defense, education, health, public parks, and infrastructure.

4.2 Environmental Front

Environmental regulations are critical institutional mechanisms governments use to sustainably manage natural resources and protect ecosystems. These regulations are designed to conserve resources, reduce pollution, and ensure that economic development does not compromise the environment's health and resilience (Alexander, 2006). A brief overview of how environmental regulations contribute to conservation and the efficient use of resources is provided below:

Environmental regulations limit emissions of pollutants from industries, vehicles, agricultural activities, and other sources, reducing air and water pollution and protecting public health. Land use, water rights, and natural resource management laws ensure that these resources are used sustainably, preserving them for future generations. Regulations like carbon pricing, emission trading systems, and renewable energy incentives are designed to reduce greenhouse gas emissions and promote the transition to a low-carbon economy. Environmental laws protect endangered species and ecosystems, ensuring biodiversity conservation through measures like protected areas, wildlife sanctuaries, and restrictions on hunting and logging (Waring et al., 2017).

Natural resource management (NRM) institutions have been designed over the years to sustain natural resources at the local levels, such as forests, water bodies, fisheries, and grazing lands (Behera & Engel, 2006; Behera & Mishra, 2018; Carney & Farrington, 2005; Das et al., 2020). These institutions are designed to govern and regulate the use of natural resources to ensure their long-term viability, prevent over-exploitation, and balance the needs of various stakeholders. Prominent scholars who have contributed significantly to the understanding of NRM institutions are Elinor Ostrom, the Nobel Laurent from Indiana University, US, and other scholars, both from the West and East. Elinor Ostrom's work on common-pool resources (CPRs) is foundational in the field of NRM. Her research challenged the conventional wisdom of the "tragedy of the commons," which argued that individuals acting in their self-interest would inevitably over-exploit common resources, leading to resource depletion (Ostrom, 1990). Ostrom argued that local communities could effectively manage their resources through collective action, provided they developed appropriate institutions.

Ostrom identified eight design principles that successful CPR institutions often follow (Behera, 2008). These include clearly defined boundaries, congruence between appropriation and provision rules and local conditions, collective-choice arrangements, monitoring, graduated sanctions, conflict-resolution mechanisms, minimal recognition of rights to organize, and nested enterprises for larger systems. Ostrom advocated for governance systems where multiple and overlapping authorities at different scales manage resources. This contrasts with monolithic, top-down approaches, and it allows for flexibility, adaptability, and responsiveness to local conditions. Through extensive fieldwork and case studies, Ostrom demonstrated that local communities could and did manage resources sustainably. Her work included studies on irrigation systems in Nepal, fisheries in Turkey, and forest resources in India and Nepal.

Ostrom's work showed that local communities could be effective stewards of resources, countering the belief that state or market mechanisms are the only solutions. Her research highlighted the importance of understanding local contexts and the potential for diverse institutional arrangements to manage resources sustainably. Although Ostrom gathered sufficient case studies from across the world in support of her arguments, however, her institutional analysis approach is not free from limitations. While Ostrom's principles are effective for small to medium-sized communities, scaling them up to manage larger, more complex systems remains a challenge (Behera, 2009). The diversity of institutions and the complexity of polycentric governance can make implementation difficult, especially in regions with weak governance or external pressures (Behera & Engel, 2006).

4.3 Integration of Economic and Environmental Regulations

Human societies have attempted to integrate both economic and environmental regulations to achieve sustainable development (WECD, 1987). Policies promoting the green economy integrate economic growth with environmental sustainability (Georgeson et al., 2017). Incentives for clean technology, green jobs, and sustainable business practices help solve economic problems such as unemployment and poverty while reducing environmental impact significantly. Regulations requiring businesses to report on environmental impacts and adhere to sustainability standards ensure that economic activities do not harm the environment, leading to more responsible corporate behaviour in the form of aggressive adoption of corporate social and environmental responsibilities by countries across the globe (Latapi et al., 2019). By establishing clear rules, enforcing compliance, and creating incentives for sustainable practices, institutions play a vital role in ensuring that economic development does not come at the expense of the environment and that both are managed in a way that benefits society as a whole.

4.4 Understanding the Failures of Current Institutional Arrangements

As outlined above, although the human system has tried to design elaborate institutional mechanisms in terms of laws and regulations to sustainably govern both the economic and environmental systems, these institutional measures did not succeed in effectively solving both systems. Hence, it is imperative to critically examine the market system, which involves its structures, operations, and outcomes, particularly concerning economic inequality and environmental degradation.

The market system often leads to significant disparities in income and wealth distribution. Those with capital and resources tend to accumulate more wealth, while those without these critical resources fall behind, exacerbating economic inequality (Connell, 2013). The current market system facilitates capital accumulation by a small elite. This is evident in the growing wealth of billionaires compared to the stagnation or decline in the wealth of the middle and lower classes (Sachs, 2015). Executive compensation and shareholder returns are prioritized over wages for ordinary workers, leading to a significant income gap. The present market system has also produced labour market issues as well (Edwards, 1994). The rise of gig economy jobs often lacks security, benefits, and fair wages, contributing to economic instability for many workers. Technological advancements, while increasing productivity, often result in job losses, disproportionately affecting low-skilled workers (King et al., 2017). Quality education and healthcare are often inaccessible to lower-income individuals, perpetuating the cycle of poverty. The affordability crisis in housing markets makes it difficult for low- and middle-income families to secure stable housing.

Markets often fail to account for externalities, which are costs or benefits that affect third parties who did not choose to incur those costs or benefits (Behera & Reddy, 2002; Reddy & Behera, 2006). For example, pollution from a factory imposes health and environmental costs on the community, which are not reflected in the market prices of the factory's products (Reddy & Behera, 2006). Public goods, like clean air, national defense, and public parks, are non-excludable and non-rivalrous, meaning they cannot be confined to paying customers, and one person's use does not reduce their availability to others. The market tends to underprovide these goods because they are not profitable to produce.

The concentration of market power in the hands of a few firms can lead to monopolies or oligopolies, which can manipulate prices, reduce consumer choice, and stifle innovation (Tepper, 2018). This concentration also contributes to economic inequality. Firms with significant market power can engage in anticompetitive practices, such as predatory pricing or collusion, to maintain their dominance (Tepper, 2018). Markets operate efficiently when all participants have access to complete and accurate information. However, in reality, information is often imperfect or asymmetric, with one party having more or better information than the other. This can lead to market failures, such as adverse selection and moral hazard. Consumers may be unable to make informed decisions due to a lack of information about product quality, safety, or long-term impacts, leading to suboptimal market outcomes.

Market-driven resource extraction often prioritizes short-term gains over long-term sustainability, leading to overexploitation of natural resources and environmental degradation (Chen, 2021). Similarly, Markets do not adequately address the global challenge of climate change, as the costs of carbon emissions and other pollutants are not fully integrated into market prices (Stern & Stiglitz, 2021). Market systems can lead to the exploitation of labour, particularly in developing countries, where workers may face poor working conditions, low wages, and a lack of labour rights. Markets do not inherently provide for the social welfare of all individuals, often necessitating government intervention to ensure access to healthcare, education, and social security. Large corporations and wealthy individuals can exert disproportionate influence over political processes, leading to regulatory capture where regulations favour business interests over public welfare (Carpenter & Moss, 2013; Lindsey & Teles, 2017). Effective regulation to correct market failures can be hindered by lobbying and political pressure from powerful market players (ibids).

5 Political and Economic Systems

Political and economic systems are deeply intertwined, with each influencing the development, functioning, and outcomes of the other (Steiner-Khamsi, 2010). Democratic countries typically support capitalist economies, emphasizing individual economic freedom, private property and enterprise, and free markets. Democratic governments often implement regulatory frameworks to address market failures and social safety nets to protect vulnerable populations living in the margins, balancing free market principles with social welfare considerations. Authoritarian regimes may adopt centrally planned economies where the state controls the production, distribution, and pricing of goods and services. In such systems, economic decisions are made by a centralized authority, often resulting in less economic freedom but greater control over economic outcomes.

Many countries operate under mixed economies, combining elements of capitalism and socialism. The types of political systems determine the extent and nature of government intervention in the economy. The political system influences economic policies, including the balance between market forces and state intervention. Political instability, such as frequent changes in government, corruption, or conflict, can disrupt economic activities, deter investment, and hinder economic growth. In unstable political systems, frequent policy changes can create an unpredictable business environment, affecting economic performance negatively. Economic elites and large corporations tend to exert significant influence over political systems through lobbying, campaign contributions, and other means, shaping policies to favour their interests. This influence often leads to regulatory capture, where regulatory agencies serve the interests of the industries they regulate rather than the public interest (Carpenter & Moss, 2013; Lindsey & Teles, 2017).

Political systems influence a country's stance on free trade versus protectionism. Democratic nations might advocate for free trade agreements, while others might prioritize protecting domestic industries. Political relationships and alliances shape economic agreements and trade policies, impacting global economic integration and regional economic blocs. Political systems use economic sanctions as a tool of foreign policy to exert pressure on other nations, influencing international economic relations. Diplomatic efforts often focus on securing favourable economic terms, trade deals, and investment opportunities, reflecting the political objectives of the nation. Hence, institutional failure leads to the breakdown of systems and rules designed to guide and regulate the functioning of organizations, governments, and societies. These failures often lead to negative outcomes on various levels, including economic, social and environmental. The utter mismanagement of economic and environmental systems today is largely attributed to institutional failures, which failed to structure human behaviour in the right direction.

6 Conclusions

The current economic paradigm is riddled with fundamental ailments that pose significant challenges to environmental sustainability while driving economic growth and prosperity. The pursuit of GDP growth often comes at the expense of ecological balance, leading to resource depletion, environmental degradation, and climate change. Additionally, the unequal distribution of wealth and economic benefits exacerbates social disparities, undermining the true essence of prosperity.

A paradigm shift is necessary to achieve a more sustainable future—one that integrates economic policies with environmental responsibility and social equity. Alternative economic models, such as the solidarity economy, well-being economics, circular economy and green growth strategies, offer promising solutions by promoting resource efficiency, reducing waste, and prioritizing long-term well-being over short-term gains. Governments, businesses, and individuals must work collaboratively to redefine economic success, ensuring that growth does not compromise the planet's health or future generations' ability to thrive. Ultimately, striking a balance between economic progress and ecological sustainability is imperative. By addressing the shortcomings of the current economic system and embracing sustainable practices, societies can pave the way for a resilient, inclusive, and environmentally responsible future.

While technological and institutional innovations are often heralded as solutions to the challenges of economic growth, prosperity, and environmental sustainability, they also have significant weaknesses that hinder their effectiveness in addressing the ailments of the current economic paradigm. Despite improving efficiency and reducing environmental impact in some areas, technological advancements often lead to the rebound effect, where increased efficiency results in greater resource consumption rather than conservation. Additionally, many green technologies remain costly, inaccessible, or dependent on unsustainable supply chains, limiting their large-scale adoption.

Institutional innovations, including regulatory frameworks and international agreements, also face critical shortcomings. Bureaucratic inefficiencies, lack of political will, and corporate influence often dilute environmental policies, making them ineffective or difficult to enforce. Moreover, many institutions prioritize short-term economic gains over long-term sustainability, reinforcing a growth-driven model that perpetuates environmental degradation and social inequalities.

In order to overcome these limitations, a more holistic approach is needed—one that goes beyond technological fixes and institutional reforms to address the root causes of unsustainable economic practices. Without a fundamental shift in economic priorities toward true sustainability, technological and institutional innovations alone will remain insufficient in solving the deep-seated challenges of the present economic paradigm.

References

Achilladelis, B., Schwarzkopf, A., & Cines, M. (1990). The dynamics of technological innovation: The case of the chemical industry. *Research Policy, 19*(1), 1–34.

Aksoy, Z. (2016). Biodiversity and biotechnology in the agriculture sector. In *Environmentalism in Turkey* (pp. 249–262). Routledge.

Alegre, J., & Pasamar, S. (2018). Firm innovativeness and work-life balance. *Technology Analysis & Strategic Management, 30*(4), 421–433.

Alexander, E. (2006). Institutional design for sustainable development. *Town Planning Review, 77*(1), 1–27.

Aydalot, P., & Keeble, D. (2018). High-technology industry and innovative environments in Europe: An overview. *High Technology Industry and Innovative Environments*, 1–21.

Ayim, C., Kassahun, A., Addison, C., & Tekinerdogan, B. (2022). Adoption of ICT innovations in the agriculture sector in Africa: A review of the literature. *Agriculture & Food Security, 11*(1), 22.

Behera, B. (2008). Institutional dynamics and natural resource management: A study of JFM in Andhra Pradesh. *Journal of Rural Development, 27*(4): 575–606

Behera, B. (2009). Explaining the performance of state-community joint forest management in India. *Ecological Economics, 69*(1), 177–185.

Behera, B. (2012). *Consumption behaviour, economic growth, and sustainable well-being.* Paper presented in international conference "Planet Under Pressure: New Knowledge Towards Solution", held in London, United Kingdom, 26–29 March. https://elsevier.conferenceservices.net/reports/template/onetextabstract.xml?xsl=template/onetextabstract.xsl&conferenceID=2808&abstractID=567747

Behera, B. (2013). *Can changing climate be reversed and/or halted without radically transforming the human psyche? Transformation in a changing climate.* In International Conference in Oslo 19–21 June, Conference Proceedings (p. 282). http://cchange.no/wp-content/uploads/2014/02/Proceedings-Transformation-in-a-changing-climate_interactive.pdf

Behera, B. (2017). *Can sustainability problem be addressed without radically transforming the human psyche?* Paper presented in a seminar organised by the Institute of Geography and Sustainability (IGD), University of Laussane, Switzerland, 12 June (13.30–14.30) at Geopolis 3799. https://agenda.unil.ch/display?id=1496411660716

Behera, B. (2023). A new economy for happiness. In: S. Chetri, T. Dutta, M. K. Mandal, P. Patnaik, (Eds.), *Handbook of happiness.* Springer. https://doi.org/10.1007/978-981-99-2637-4_8

Behera, B., & Engel, S. (2006). Institutional analysis of evolution of joint forest management (JFM) in India: A new institutional economics (NIE) approach. *Forest Policy and Economics, 8*(4): 350–362.

Behera, B., Mishra, P. (2007). Acceleration of agricultural growth in India: Suggestive policy framework. *Economic and Political Weekly 42*(42).

Behera, B., & Mishra, P. (2018). Democratic local institutions for sustainable management and use of minor irrigation systems: Experience of Pani Panchayats in Odisha, India. *Water Economics and Policy, 4*(3), 1–27.

Behera, B., & Reddy, V. R. (2002). Environment and accountability: Impact of industrial pollution on rural communities. *Economic and Political Weekly, 36*(3).

Basheer, M., Nechifor, V., Calzadilla, A., Ringler, C., Hulme, D., & Harou, J. J. (2022). Balancing national economic policy outcomes for sustainable development. *Nature Communication, 13*, 5041. https://doi.org/10.1038/s41467-022-32415-9

Burger, J. R. (2018). Modelling humanity's predicament. *Nature Sustainability, 1*(1), 15–16.

Cafaro, P., Hansson, P., & Götmark, F. (2022). Overpopulation is a major cause of biodiversity loss and smaller human populations are necessary to preserve what is left. *Biological Conservation, 272*. https://doi.org/10.1016/j.biocon.2022.109646

Carney, D., & Farrington, J. (2005). *Natural resource management and institutional change*. Routledge.

Carpenter, D., & Moss, D. A. (Eds.). (2013). *Preventing regulatory capture: Special interest influence and how to limit it*. Cambridge University Press.

Chen, Y. (2021). Addressing environmental challenges through circular economy models. *Journal of Energy and Environmental Policy Options, 4*(3), 9–15.

Crist, E. (2022). Witnessing mass extinction: What's invisible, what's visible, what's possible. *Biological Conservation, 275*. https://doi.org/10.1016/j.biocon.2022.109696

Connell, R. (2013). Why do market 'reforms' persistently increase inequality? *Discourse: Studies in the Cultural Politics of Education, 34*(2), 279–285.

Das, S., Behera, B., & Mishra, A. (2020). Property rights and institutional arrangements of a Man-Made Wetland in Dryland Area of West Bengal, India. *Wetlands, 40*, 2553–2560. https://doi.org/10.1007/s13157-020-01360-y

de Zwart, M., & Stephens, D. (2019). The space (innovation) race: The inevitable relationship between military technology and innovation. *Social Science Research Network*. https://papers.ssrn.com/sol3/papers.cfm?abstract_id=3446451

Ebner, A. (2008). Institutional evolution and the political economy of governance. In *The institutions of the market: Organizations, social systems, and governance* (pp. 287–308). Oxford University Press.

Edwards, P. (1994). Economic theory, the labour market and inequality. *Work, Employment and Society, 8*(2), 273–287.

Fountas, S., Carli, G., Sørensen, C. G., Tsiropoulos, Z., Cavalaris, C., Vatsanidou, A., Liakos, B., Canavari, M., Wiebensohn, J., & Tisserye, B. A. (2015). Farm management information systems: Current situation and future perspectives. *Computers and Electronics in Agriculture, 115*, 40–50.

Ganguly, A., Chattaraj, M., Chattaraj, S., Mitra, D., Ganguly, M., Kirtania, D. K., & Konar, R. (2025). Employing technological innovation for eco-friendly waste disposal at a global scale. *Advances in Environmental Engineering and Green Technologies Book Series*, 135–152. https://doi.org/10.4018/979-8-3693-7282-1.ch008

Georgeson, L., Maslin, M., & Poessinouw, M. (2017). The global green economy: A review of concepts, definitions, measurement methodologies and their interactions. *Geo: Geography and Environment, 4*(1), e00036.

Giampietro, M., & Mayumi, K. (2018). Unraveling the complexity of the Jevons Paradox: The link between innovation, efficiency, and sustainability. *Frontiers in Energy Research, 6*, Article 349753.

Gough, I. (1994). Economic institutions and the satisfaction of human needs. *Journal of Economic Issues, 28*(1), 25–66.

Hall, B. H., & Helmers, C. (2024). *Innovation and Economic Growth*. 187–216. https://doi.org/10.1093/oso/9780197630914.003.0009

Harahap, M. A. K., Wurarah, R. N., Fathurohman, A., Suroso, A., & Iskandar, Y. (2023). Globalization substance and industrial revolution 4.0 and the role of technological innovation for economic development towards entrepreneurship. *Jurnal Bisnisman: Riset Bisnis Dan Manajemen, 4*(3), 37–51.

Hussen, M. A. (2004). *Principles of environmental economics*. Routledge.

Jackson, T. (2009). *Prosperity without growth: Economics for finite planet*. Earthscan.

Jena, L. K., & Behera, B. (2017). Environmental crisis and human wellbeing: A review. *International Journal of Development and Sustainability, 6*(8), 561–574.

Kalantari, F., Tahir, O. M., Joni, R. A., & Fatemi, E. (2018). Opportunities and challenges in sustainability of vertical farming: A review. *Journal of Landscape Ecology, 11*(1), 35–60.

King, B. A., Hammond, T., & Harrington, J. (2017). Disruptive technology: Economic consequences of artificial intelligence and the robotics revolution. *Journal of Strategic Innovation and Sustainability, 12*(2), 53–67.

Latapí Agudelo, M. A., Jóhannsdóttir, L., & Davídsdóttir, B. (2019). A literature review of the history and evolution of corporate social responsibility. *International Journal of Corporate Social Responsibility, 4*(1), 1–23.

Lindsey, B., & Teles, S. M. (2017). *The captured economy: How the powerful enrich themselves, slow down growth, and increase inequality*. Oxford University Press.

Loasby, B. J. (2001). *Market institutions and economic evolution*. In Capitalism and democracy in the 21st century: Proceedings of the International Joseph A. Schumpeter Society Conference, Vienna 1998 "Capitalism and Socialism in the 21st Century" (pp. 303–315). Physica-Verlag HD

Maclaurin, W. R. (1950). The process of technological innovation: The launching of a new scientific industry. *The American Economic Review, 40*(1), 90–112. http://www.jstor.org/stable/1802812

Meadows, D. H., Meadows, D. L., Randers, J., & Behrens, W. W. (1972). *The limits to growth: A report for The Club of Rome's project on the predicament of mankind*. Universe Books.

Méndez-Zambrano, P. V., Tierra Pérez, L. P., Ureta Valdez, R. E., & Flores Orozco, Á. P. (2023). Technological innovations for agricultural production from an environmental perspective: A review. *Sustainability, 15*(22), 16100.

Morrar, R., Arman, H., & Mousa, S. (2017). The fourth industrial revolution (Industry 4.0): A social innovation perspective. *Technology Innovation Management Review, 7*(11), 12–20.

Mukherjee, S., Mishra, A. K., Ashfaq, M., & Kao, S-C. (2022). Relative effect of anthropogenic warming and natural climate variability to changes in Compound drought and heatwaves. *Journal of Hydrology, 605*. https://doi.org/10.1016/j.jhydrol.2021.127396

Neyapti, B. (2013). Modeling institutional evolution. *Economic Systems, 37*(1), 1–16.

North, D. C. (2003). The role of institutions in economic development. UN.

Osborne, E., Hu, X., Hall, E. R., et al. (2022). Ocean acidification in the Gulf of Mexico: Drivers, impacts, and unknowns. *Progress in Oceanography,* 209. https://doi.org/10.1016/j.pocean.2022.102882

Ostrom, E. (1990). *Governing the commons: The evolution of institutions for collective action*. Cambridge University Press.

Pang, T. Y. (2024). Digital economy, technological innovation, and environmental pollution in different regions: An analysis based on a threshold model. *Deleted Journal, 6*(1), 76–84. https://doi.org/10.54254/2977-5701/6/2024051

Pantanella, E., Cardarelli, M., Colla, G., Rea, E., & Marcucci, A. (2010). Aquaponics vs. hydroponics: production and quality of lettuce crop. In *XXVIII International horticultural congress on science and horticulture for people (IHC2010): International symposium on 927* (pp. 887–893).

Perman, R., Ma, Y., McGilvray, J., & Common, M. (2003). *Natural resource and environmental economics*. Pearson Education Limited.

Polimeni, J. M., Mayumi, K., Giampietro, M., & Alcott, B. (2015). *The myth of resource efficiency: The Jevons paradox*. Routledge.
Popkova, E. G., Ragulina, Y. V., & Bogoviz, A. V. (Eds.). (2019). *Industry 4.0: Industrial revolution of the 21st century* (Vol. 169, p. 249). Springer.
Qamruzzaman, Md., & Kor, S. (2024). Navigating the path to environmental sustainability: Insights from CIVETS on the intersection of ICT diffusion, natural resources, and green technological innovation. *PLOS ONE, 19*(12), e0309264. https://doi.org/10.1371/journal.pone.0309264
Rausser, G., Choi, E., & Bayen, A. (2023). Public-private partnerships in fostering outer space innovations. *Proceedings of the National Academy of Sciences of the United States of America, 120*(43), e2222013120. https://doi.org/10.1073/pnas.2222013120
Reddy, V. R., & Behera, B. (2006). Impact of water pollution on rural communities: An economic analysis. *Ecological Economics 58*(3), 520–537.
Sachs, J. (2015). *The age of sustainable development*. Columbia University Press.
Sayara, T., Amarneh, B., Saleh, T., Aslan, K., Abuhanish, R., & Jawabreh, A. (2016). Hydroponic and aquaponic systems for sustainable agriculture and environment. *International Journal of Plant Science and Ecology, 2*(3), 23–29.
Sharma, A. (2024). Role of technology innovation in economic growth: A bibliometric analysis. *International Journal of Experimental Research and Review, 44*, 121–135. https://doi.org/10.52756/ijerr.2024.v44spl.011
Steiner-Khamsi, G. (2010). The politics and economics of comparison. *Comparative Education Review, 54*(3), 323–342.
Stern, N., & Stiglitz, J. E. (2021). *The social cost of carbon, risk, distribution, market failures: An alternative approach* (vol. 15). National Bureau of Economic Research.
Tabash, M. I., Farooq, U., Hassen, M., & El Refae, G. A. (2025). Do technological innovation and financial development determine environmental quality? Empirical evidence from Arab countries. *Review of Accounting and Finance*. https://doi.org/10.1108/raf-02-2024-0050
Tepper, J. (2018). *The myth of capitalism: Monopolies and the death of competition*. John Wiley & Sons.
UNDP. (1990). *Human development report 1990*. Oxford University Press.
Waring, T. M., Goff, S. H., & Smaldino, P. E. (2017). The coevolution of economic institutions and sustainable consumption via cultural group selection. *Ecological Economics, 131*, 524–532.
Wegner, G. (2003). Evolutionary markets and the design of institutional policy. In *The evolutionary analysis of economic policy* (pp. 46–66). Edward Elgar Publishing.
WECD (1987). From one earth to one world: an overview. In *World commission on environment and development (WECD)*, Oxford University Press.
York, R., & McGee, J. A. (2016). Understanding the Jevons paradox. *Environmental Sociology, 2*(1), 77–87.
Zhang, H. (2024). The impact of technological innovation on economic growth: A study based on relevant cases. *Advances in Economics, Management and Political Sciences, 139*(1), 130–136. https://doi.org/10.54254/2754-1169/2024.19289
Zhang, R., Liu, J., Li, Y., & Guo, X. (2025). The role of technological innovation and institutional quality in environmental and economic growth sustainability in emerging Asian countries. *Frontiers in Environmental Science, 12*. https://doi.org/10.3389/fenvs.2024.1510120

Chapter 3
Post-Growth Approaches and Well-Being: A Critique

1 Introduction

In order to overcome the pitfalls and ailments of the current growth-centric paradigm (as discussed in Chap. 2), some interesting de-growth or post-growth approaches have been advanced. This chapter provides a detailed account of some of these approaches and critically examines the feasibility of adopting these approaches as development models. GDP-centred growth approaches have been attracting a lot of flak, especially after the 1990s when most of the developing countries embarked on economic reforms. The purpose and objective of these reforms was to enhance growth performance and catch up with the 'so-called' developed economies. While some developing countries could reduce their poverty burden, they could not sustain the trends. Though most of these countries could address absolute poverty to some extent, economic inequalities have increased substantially. This has made them hungrier for growth (GDP), which is not sustained for longer periods. There are a number of reasons for unsustainable growth and the most important are the resource and environmental degradation associated with high growth rates. In some regions and in the case of specific resources the degradation has been reaching unsustainable levels. Another important factor is that even the developed countries have been experiencing an economic slowdown that has impacted the developing economies. Besides, the cost escalation in the developed world has increased their reliance on imports. While the increased import demand from the developed countries has benefited a few countries like China, the benefits have been associated with huge environmental costs. Greater reliance on imports has pushed the unemployment rates up in the developed countries, setting the downward swirl of economic prosperity.

As a result, most developing countries have been missing the timelines in achieving even the most important sustainable development goals like poverty, water and sanitation, health care, etc. That is, achieving sustainable development goals has become a mirage for most of these countries and some of the developed countries are staring at the prospect of slipping from their existing levels. These economic tensions,

V. R. Reddy and B. Behera, *Understanding the Current Global Socio-economic and Ecological Crises*, https://doi.org/10.1007/978-3-031-98600-0_3

coupled with increasing food insecurity associated with climate risks, natural calamities (COVID-19) and political instability across the countries, have given rise to alternative growth approaches or post-growth theories. Policy makers and international organisations have started looking at these alternatives. Even hard-core free-market protagonists have started doubting the sustainability of growth-centred economic policies and strategies and started exploring alternative approaches. Most important among these include 'Needs-Based', 'Rights-Based/'Human Rights-Based'/'Right to Livelihoods' and 'Decoupling Growth' approaches. The Needs-Based approach suggests that production should be focused on specific goods and services that are necessary to improve capabilities and meet human needs at a high standard, while ensuring universal access to key goods and services through public provisioning and decommodification. Human Rights-Based approach to development is an approach that advocates protecting human rights through the provision of legal sanctity to basic needs i.e., development policies should be guided by principles of human rights. This approach has been developed and adopted widely by the UN agencies. Besides, the number of international NGOs like Oxfam and development agencies like GIZ, EU, etc., have been adopting this approach. The decoupling growth approach talks about 'growth decoupling' (GDP growing faster than GHG emissions) and 'wellbeing decoupling' (augmented human development growing faster than GDP). It is argued that the decoupling process would reduce the ecological footprint and pave the way for sustainable development (Juan et al., 2024).

These alternative approaches are different from the standard growth-centred development models that have basic economic principles at their core. While growth-centred approaches can be managed and maneuvered with policy adjustments like interest rates, trade tariffs, incentives, protection, etc., post-growth models require fundamental changes that go against the basic economic principles like selfish motive, appropriation, enclosure, commodification and free markets. Thus, the growth-centred approaches are easy to manage through macro adjustments at the macro level, alternative approaches require micro-economic adjustments like behavioural changes at the individual level. Reducing the demand for luxury goods, increasing the production of goods that are needed by the poor, equitable distribution of natural resources, etc., may not be possible through policy adjustments. That is, the 'transformation of the human mind' is the key to bringing changes in ecological, economic and social systems (Krishnamurti, 2010). In other words, the human psyche should move towards inner (non-material) development and away from external (material) development (Thurman, 2012). This calls for changes in basic understanding of development and welfare, that need to be imbibed in the knowledge base, learning and education systems. In this chapter, we examine the basic premise of these approaches and their potential for achieving sustainable development or checking damaging growth. Besides, the applicability of these approaches makes them effective at the policy level.

2 Needs-Based Approach

Needs-based approach (Hickel & Sullivan, 2024) is persuasive in suggesting that global poverty can be effectively minimised or removed by simply altering what is being produced. Using the example of China, it is argued that the high GDP growth rates have, in fact, increased poverty when compared to the earlier low growth period associated with state provisioning. This is mainly associated with the increased cost of basic needs in developing countries after the advent of economic reforms (high GDP growth regime). However, the China example of GDP growth and the extent of poverty during pre-growth and growth periods is misplaced and misleading. For, the availability and validity of data during the cultural revolution in China are quite uncertain. One can present an opposite picture with the Indian growth experience. India has experienced a substantial decline in poverty after the 1990s the period associated with high GDP growth. Though India experienced increased inequalities, discarding the 'trickle down' effects of growth, the argument on GDP-poverty linkages, based on the experience of China, is rather naive. However, the GDP growth in China as well as India has come at huge environmental and resource degradation costs. For instance, China's annual share of natural resources costs to total costs was more than 80%, reflecting the high dependence on resources and their degradation (Xiang et al., 2021).

According to the needs-based approach the focus of production should shift towards basic needs that can address extreme poverty and higher-level needs (decent housing, sanitation, hygiene, health care, education, etc.) that can ensure decent living standards (DLS). The underlying premise for this is that it takes four times the present growth rate to bring all the countries to a decent living standard i.e., US$ 30 per capita/day. This would, in turn, impose huge environmental burden, leading to destructive development. With shifts in production away from luxury goods the present growth can support the DLS for the global population. This means that about 20% of the world's population should cut their present consumption patterns drastically and deprive the aspirations of attaining these life-styles of the majority population's (middle class). And the producers of luxury goods should shut/minimise their production, which includes mining high-value minerals like gold, diamonds, uranium, etc. This also means loss of employment and a substantial reduction in defense spending. It may be noted that a number of developed countries, including the USA, thrive on defense exports.[1]

The million-dollar question is how to persuade the majority of the global population to compromise on their present consumption levels and their future aspirations for material goods. How practical is it to persuade the producers to reduce the production of luxury goods that provide high-profit margins? More importantly, the political economy factors that support the rich lobby groups, viz., corporates and farmer groups. Moreover, the transition time lag for the shift in production strategies (luxury goods to basic needs) could be painful for the vulnerable sections, and hence,

[1] Of late even countries like Japan are enhancing their defense spending. It is also true that some of the developing countries survive on drug trade.

political entrepreneurs may not be keen to support such a transition. This is more so in the context of democratic systems. Even China, under a socialist community regime, had to move away from the ideology of basic goods provision with equal access in favour of luxury goods. This calls for hard-core communist or dictatorial regimes to shift towards basic goods. At the same time, the benevolence and efficiency of these governance regimes have been disproved in multiple cases over the years. These regimes proved totalitarian and failed to ensure basic human rights and well-being.

Another important aspect that could help realise the objective of equitable DLS is through the public provision of basic goods, as argued by Hickel and Sullivan (2024) based on a comparative study of low-growth and high-growth periods. The experience of China has clearly indicated that provisioning is not sustainable, as the approach failed to move towards equitable DLS, though it might have helped in reducing extreme poverty. That is the economy failed to generate enough resources to move the populations to higher living standards (DLS). Using the number of people getting out of extreme poverty may not reflect the overall well-being of the population, i.e., the majority of the population (middle class) may be doing well during the high growth periods despite the rise in extreme poverty. The experience of dominant state provisioning in countries like India has amply demonstrated the inefficiencies as well as the sticky nature of poverty during low-growth regimes. On the contrary, India experienced a rapid decline in extreme poverty during the high growth period (after the 1990s).

In this context, Stiglitz (2024), in his recent book "Road to Freedom: Economics and Society", puts forth a more balanced approach to increasing the role of state provisioning while maintaining the status-quo in production structure. Financing the state provisioning is a major challenge. That is, State provisioning can be funded through a high growth trajectory, which comes with high resource degradation. This squarely puts us back to development with degradation. In this approach, the trade-off between the extent of public provision and resource degradation is likely to continue unabated. While Stiglitz's approach focuses on reduced poverty with better equity, it doesn't entirely subscribe to de-growth or post-growth approaches. This could be feasible if the trade-off between growth and resource degradation is kept at moderate levels. Though this is not impossible to achieve, this may go against the aspirations of the majority populations (i.e., people aspiring to move beyond basic needs and DLS). This could be a potential threat to the feasibility of this approach.

The 'needs-based approach' on paper looks like an attractive solution to address global poverty effectively, but a deeper probe into its implementation exposes a number of deep-rooted behavioural and practical constraints. It is based on some interesting arithmetic calculations without any theory to support it. In the absence of a strong theoretical basis and policy approach it remains as one more idealistic approach like socialist/communist ideologies that were embraced during the 1960s and 1970s to address human misery and inequalities. It doesn't acknowledge the natural laws of some level of inequity in income and well-being that are associated with human capabilities and endowments. Even the Marxist philosophy with strong theoretical underpinning failed to overcome the natural human instincts. In fact, these natural instincts and economic principles like selfish motives and non-declining utility for money or wealth have been fuelling human prosperity over centuries. Even the centuries-old ancient scriptures of different civilisations are not devoid of discerptions on inequities, i. e., intra-community differences in wealth, lifestyles and consumption patterns. While the luxury lifestyles of certain communities or individuals were accepted, the deprivation or extreme poverty of vulnerable groups was seen as usurpation and poor governance. One finds a plethora of examples of good governance and less greed or benevolence among the ruling gentry that help ameliorate extreme poverty and misery, rather than changing the production structure. And it is hard to find any evidence of absolute equity even in the Indian epic of '*Ramayana*' or '*Rama Rajya*' the epitome of good governance.

In the present context, there is ample evidence that self-interest is taking extreme forms i. e., excessive and increasing margins of the corporate sector and corruption among the ruling classes. In recent years corporate profits have been growing at a much greater pace when compared to per capita GDP. For, suppression of wages is one of the main reasons for rising corporate profits. The contrast is much higher when compared to the farming sector which accommodates 50% of the population in the developing countries. The spike in corporate profits is contributing substantially to inflationary trends. The vicious circle of rising corporate profits and inflationary trends is severely denting the living standards of the poor and middle class. This, in turn, is resulting in a sharp rise in inequalities. This is mainly attributed to the nexus between the ruling gentry (politicians and administrators) and the corporate sector. At the same time, the farming-dependent rural communities accounting for about 50% of the total population in the developing world are facing deep economic crisis over the decades. These farm communities are increasingly becoming unviable and surviving on increasing debts. And non-farming or land-less population are eking out their livelihoods through wage employment.

Most of these communities can't afford decent living standards as defined by the 'needs-based approach'. Despite the fact that the majority of them depend on public provisioning hardly a fraction of the public spending meant for them really reaches them. This is well reflected in the famous quote from one of the Indian Prime ministers, Mr. Rajiv Gandhi: "*Only 15 paisa of a rupee (100 paisa) reaches the hands of the beneficiaries*". For the remaining 90 paisa goes into pilferage and corruption. Corruption is not only spreading but also growing at a faster pace in recent decades (often resulting in the accumulation of black money). Over the years the extent and magnitude of this component has gone up substantially with increased public provisioning. For, populism has increased many folds in most developing countries. This is mainly attributed to unsaturated human greed, which is an extreme form of self-interest. This form of greed is benefiting a minority at the cost of the majority and hence the spiralling inequalities in recent years. As aptly put by Mahatma Gandhi "*Earth provides enough for everybody's need but not enough for anybody's greed*". According to Gandhi nature is the source of life and not merely a resource for the economy. In the present context nature or natural resources have become the source of economic growth. There is need for looking at the real root causes for the existing state of affairs rather than attributing to some esoteric reasons and unpractical solutions like altering the consumption and production priorities. At the same time, looking at GDP growth as the main culprit amounts to throwing the baby with bath water. Besides, there is a need for inventing and focusing on environmentally friendly growth processes.

3 Decoupling Growth

One of the main challenges of development in the twenty-first century is the promotion of environmentally friendly growth. The Industrial Revolution triggered modern economic growth that facilitated the accumulation of wealth at an increasing rate of inequalities. This has led to unprecedented environmental impacts which threatened the material prosperity itself. It is argued that the ongoing environmental destruction associated with the accumulation of wealth could be reversed or minimized by decoupling economic growth from its environmental degradation and impacts, as proposed in green growth models. However, the feasibility of such a growth and sustaining it is a major challenge. An alternative would be decoupling wellbeing from economic growth as proposed in post-growth strategies. This kind of decoupling brings us back to 'needs-based' development. The feasibility of such a welfare centric development has been discussed in the previous section.

Historical analysis of global growth and development has provided some evidence on decoupling patterns during certain periods and certain regions (Juan et al., 2024). The long-term analysis suggests that economies at lower-level equilibriums have managed to decouple the augmented human development index (AHDI) from GDP per capita. That is, at low income levels, some countries have experienced 'well-being de-coupling' i.e., AHDI is growing faster than GDP per capita. Growth decoupling i.e., GDP per capita grows faster than GHG emissions, takes place as the income activity levels increase (countries getting richer). But these episodes of decoupling are neither sustained nor irreversible (Juan et al., 2024). The historical evidence about decoupling appears to be incidental as a part of the economic cycle, as no country in the world could manage to achieve high levels of well-being within the ecological boundaries. More importantly, these decoupling episodes over the years is clearly empirical and random in the countries, as there is no robust theoretical explanation associated with the patterns.

The feasibility of decoupling is questioned in a study of China where the relationship between growth and environmental impacts across provinces was analyzed using the data between 2000 and 2015 (Yefeng et al., 2022). The extent of coupling between economic growth and environmental impact (measured in terms of progress in sustainable development goals) ranges between 52 and 83% across provinces of China. Though there has been a decline in the extent of coupling over the years, this clearly indicates that decoupling growth from environmental impacts is neither easy nor automatic. The extent of growth de-coupling or coupling could increase or decrease over the years at the provincial or country level, but the linkages between GDP per capita and resource degradation are ever-increasing at the global level. This is mainly due to the shifts in production bases, especially of the polluting goods and services. The increase in the extent of decoupling among the high-income countries is due to the shift of polluting technologies from developed to developing countries. This process, while reducing the resource degradation in the high-income countries, it increases the degradation in the low-income countries. Overall, the extent of coupling between GDP and resource degradation is increasing at the global level reducing the future potential of well-being.

Decoupling growth and resource degradation are critical for achieving sustainable development that takes care of future needs. Given the inter-linkages and externalities associated with any of the natural resource systems as well as greenhouse gas or carbon emissions across the geographical contexts, the decoupling needs to be aimed and achieved at the global level rather than the country or provincial level. For, resource degradation in one place/country would have far and wide impacts over time and across geographies. In some cases, the impacts would be more severe in the non-contributing areas/countries viz., flooding, water stress, air pollution, etc. For instance, the construction of dams upstream may reduce flooding down-stream but may cause water stress as well as increase the future flood risks, stubble burning by Haryana farmers impacts Delhi's air quality more and the use of wood for winter heating in the UK used to impact the air quality in the neighbouring countries due to climatic conditions. The same is true with the shifting of polluting technologies from developed to developing countries.

Thus, effectively addressing decoupling at scale is a long-term process involving multiple dimensions like behavioural, social, economic, philosophical, technological, etc. While number of examples and instances are reported where such changes can be managed and made decoupling possible. But these are achieved at a very micro-level with little or no impact on the macro developments. More importantly, sustainability, scaling-up and scaling-out of these practices are proving to be difficult, if not impossible. In most cases, the success of decoupling is linked to few individuals and their leadership and in some cases, it is linked to external support systems, viz., funding, technologies, institutions, etc. This is mainly due to the reason that the core of human behaviour of self-interest and accumulating nature remains unchanged. The temporary adjustments to these core values are mainly due to the changing socio-economic circumstances. Besides, these cases take place only in the rural settings and farming communities of the developing world. These communities act and adopt sustainable practices when faced with extreme conditions, especially with natural resources like water, forests, etc. It is rare to find such cases in the urban communities or the developed countries. The experience with this model across the world is that it becomes ineffective once the leadership or support systems are weakened or removed. This is mainly due to the reason that the institutions that are critical for the success of these case studies are also often dependent on local leadership. In the absence of appropriate leadership, even the external support is not effective. Unfortunately, unlike technologies, leadership cannot be multiplied or cloned and scaled up.

Institutions are crucial for the organisation of production processes and influencing consumption patterns. Participatory philosophy is central to putting sustainable and equitable processes in place that ensure the effectiveness and sustenance of these institutions. This is more so in the context of monopolistic markets where organising numerous producers and consumers is required. At the policy level, participatory or collective approaches and strategies are being treated and promoted as a technology rather than a philosophy. The philosophy of understanding the importance of the common good and community sharing needs to be imbibed from the very early stages of human development. In this context, the basic and primary education system plays a crucial role in transforming the human mind. The present education system is individual centric with self-interest and competition-focused. This, in turn, promotes acquisitiveness and accumulation rather than distribution and sharing of common goods. Even the Asian countries that have much longer ancient cultural heritage been moving towards western education system. The entire education system needs to be oriented towards such a behavioural change. A more detailed discussion regarding the transformative nature of education is taken up later.

4 Human Rights-Based Approach to Development (HRBAD)

The rights-based approach or human rights-based approach provides legal sanctity for ensuring human rights in a development context, i.e., provision of basic needs to the vulnerable sections. Right to Development was adopted by the UN General Assembly in 1986 and reaffirmed by the World Conference on Human Rights in 1993. Given the inter-linkages between democracy, human rights and sustainable development, these broad concepts are being supported and promoted by the UN bodies like UNDP and UNICEF (Broberg & Sano, 2017). During the 1990s, human rights theory and practice were embedded within International NGOs and termed HRBAD. The right to development is where rights to various basic needs are claimed by the people of a country. These rights are collectively acquired for the benefit of individuals. The responsibility of ensuring these rights lies with the individual states and international community by co-operating in trade, debt, finance, technology transfer and development assistance (Sengupta, 2004).

HRBAD transforms the role of the state from a benevolent provider to a subservient benefactor of ensuring and upholding the legal sanctity of human rights. While this transformation is termed as new and distinctly different from other approaches by some, others term it as new jargon without any substance (Cornwall & Nyamu-Musembi, 2004). At the same time, HRBAD increases the responsibility of the State in the provision of basic needs (goods and services). This has prompted international agencies like the World Bank to mainstream HRBAD in order to enhance the efficiency of development aid. Following the international requirements number of developing countries have adopted the rights-based approach by enacting number of policies viz., 'right to food', 'right to education', 'right to work', 'right to health', 'right to information', etc. But very few of these policies are put into practice. For, the HRBAD approach appears more idealistic and places the implementing governance structures at a high moral ground rather than bringing any fundamental changes in the way development is currently pursued. In fact, it has become part of the political discourse and is often found in the political manifestos rather than percolating through policy framework to the vulnerable sections on whose name the rights have been fought. As a result, it has remained as a slogan (political or developmental) without altering the power structure and redistribution of resources across the developing and developed worlds (Uvin, 2007). It is argued that, in a way, HRBAD has re-politicised the development discourse (Gready, 2008). But re-politicisation of development has been happening in the reverse direction i.e., political entrepreneurs using HRBAD for their own benefit rather than devolving it to the needy people.

One of the main reasons for its esoteric survival is the impracticality of its adoption and implementation in its totality. That is, the provision of rights in all spears (fields related to development) and across all the rungs of the ladder (developing/developed nations/international agencies). Adopting HRBAD at the project level is meaningless or ineffective. The realisation of the rights-based approach calls for organisational and social changes through organisational changes are a necessary precursor to social change (Vandenhole & Gready, 2014). As is the case with all ethically desirable objectives, promoting or preaching them is easier than adopting and practicing. Organisations or even nations that are seeking to promote HRBAD should first ensure that their internal management and decision-making procedures are transparent, accountable, non-discriminatory and non-exclusionary. That is, the promotion of HRBAD should start with oneself through dialogue, support and principled communication (Uvin, 2007). Aid agencies must be willing to apply the rights agenda to all of their own actions (the inward focus), and the global political economy of inequality. Pushing ethical and idealistic development practices through conditional aid doesn't work in the long run. This calls for a comprehensive and complete participation of all stakeholders at all levels. It is hard to find such practices even at the highest level such as the United Nations and its various wings that are actively involved in promoting HRBAD.

Given that the human-rights approach to development clearly requires participation, achieving participation is equally challenging as evident from the participatory development models. Ensuring participation not only requires a lot of effort and time but also situation or location-specific and also more expensive. The main challenge is that it requires a bottom-up approach. That is, the involvement of hitherto under-represented groups or communities to understand and participate in the process. At the ground level implementing agencies should provide all the relevant information that is articulated in a manner that is understood clearly. Sustaining the participation of these groups requires regular monitoring and sustaining the benefit flows. This means greater economic costs and these vulnerable groups are not in a position to share these costs. More importantly, sustained participation requires equity in terms of the distribution of costs and benefit flows. Besides, sustained participation demands transparency among donor agencies i.e., ensuring that their aims, assessments, resources, and constraints are known to all those concerned (Uvin, 2007). This, apart from ensuring wider participation, would improve the credibility of donor agencies. However, adopting such practices is more complex and difficult in the context that the adaptation of rights-based approach cannot be limited to a project.

The impact of adopting HRBAD has been marginal, as evident in the available empirical studies. While development practices adopting HRBAD varied considerably, there are no instances of a comprehensive adaptation of HRBA in totality. In most cases, the evidence is based on qualitative information. As a result, empirical evidence has been partial, viz., the right to health received greater attention than other basic needs. Though designed targeting marginalized communities with emphasis on participation and empowerment, positive impacts have been observed only at the individual and programme levels rather than at the societal and structural levels (Noh, 2021). Despite the marginal performance of HRBAD across the countries, it has some inherent value providing legal status to development goals, placing the responsibility of development on the state and building capacities of rights holders. However, the willingness of the State to enforce these human rights is effectively thwarted by the rising crony capitalism across the countries.

In a way, HRBAD has familiarity with the needs-based approach, as both aim at providing basic needs to the vulnerable communities though the process of achieving the objective is different. While the needs-based approach relies on curtailing the production of non-basic needs and consumption, HRBAD takes a legal route to achieve the objective. In both cases, achieving the objective depends on the transformation of the production systems and behavioral changes at the consumer level. Bringing changes in the legal framework and empowering vulnerable communities is a difficult task that involves high transaction costs and may go against political and corporate interests. Despite the merits of HRBAD, it is likely to remain mostly on paper and adopting them in letter and spirit is a long-drawn process. Hence, it is likely to remain a buzzword.

5 Conclusion

The review of post-growth approaches clearly brings out the ill effects of the growth-centered approach adopted across the countries so far. While post-growth models argue for equitable access to resources and entitlements to decent living standards across the sections of the population, they fail to provide any practical pathway to achieve this. Degradation of resources associated with high growth trajectories is helping the accumulation of mindless wealth in the hands of few, especially in the hands of corporations. Such a concentration, of late, is dictating the political entrepreneurs and influencing the political and policy outcomes that often go against the interests of vulnerable groups. This is very much evident in the number of countries (democratic). The recent American election appears to be the last straw in this regard. It is unfortunate that crony capitalism is spreading its tentacles wide and deep and it is a pity that it is getting wider acceptance. While this is ominous and not a welcome sign, the post-growth approaches are yet to provide practical and effective alternatives.

The 'Needs-based' approach of re-orienting the production systems towards producing basic needs goods by moving away from the production of luxury goods appears idealistic at best. This requires transformational and behavioral changes at the producer, consumer and policy maker levels. How to bring in such a massive transformation remains unresolved. Based on some incidental historical evidence, theories of decoupling growth from natural resources argue that degradation-less growth could be a possibility. But there is no clear pathway for achieving such a growth. Technology could be an option though such technologies are not available on the horizon. At the same time, such technologies need huge investments (for instance, SpaceX) and could only come through, in the present scenario, high and inequitable growth and accumulation of wealth. 'Rights-based' approaches place the discussion of post-growth theories on a different plane, as it puts the onus on the political entrepreneurs and policy makers through the provision of legal sanctity to basic human rights, viz., water, food, employment, health, education, etc. Though it tends to re-politicise the development discourse, the evidence is clear that such de-politicisation is not taking place in a desired manner. The feasibility of such an approach is questioned despite serious attempts and support from international organizations.

More importantly, none of the post-growth approaches have a clear framework that can provide consistency and pathways for achieving the sustainability objectives. These approaches have sprung from the objective of stopping undesirable growth that is destroying natural ecosystems. Such a growth is detrimental not only to future generations but also to the present generations. While these premises are very much valid, they are yet to come-up with practical solutions. This could be because each approach is focused on one aspect alone and hence fails to have a comprehensive or integrated framework to understand the problems. A more comprehensive understanding of integrating socioeconomic, ecological, behavioral, philosophical, educational and technological dimensions and their theoretical underpinnings is required in this regard. There is also a need to understand the systems in some of the countries where happiness is substituted for growth. While the post-growth theories are work in progress, they need to draw from the ancient scriptures from all the regions and religions and integrate them into the modern education system. A holistic education system could help transform the human mind and behavior. This is a long-drawn process that requires systemic changes at various levels.

References

Broberg, M., & Sano, H. O. (2017). Strengths and weaknesses in a human rights-based approach to international development—An analysis of a rights-based approach to development assistance based on practical experiences. *The International Journal of Human Rights, 22*(5), 664–680. https://doi.org/10.1080/13642987.2017.1408591

Cornwall, A., & Nyamu-Musembi, C. (2004). Putting the 'rights-based approach' to development into perspective. *Third World Quarterly, 25*(8), 1415–1437. https://doi.org/10.1080/0143659042000308447

Gready, P. (2008). Rights-based approaches to development: What is the value-added? *Development in Practice, 18*(6), 735–747. https://doi.org/10.1080/09614520802386454

Hickel, J., & Sullivan, D. (2024). How much growth is required to achieve good lives for all? Insights from needs-based analysis. *World Development Perspectives, 35*. https://doi.org/10.1016/j.wdp.2024.100612

Juan, I.-A., Travieso, T., & Aguilera, E. (2024). Unsustainable prosperity? Decoupling wellbeing, economic growth, and greenhouse gas emissions over the past 150 years. *World Development, 184*, 106754 https://doi.org/10.1016/j.worlddev.2024.106754

Krishnamurti, J. (2010). *Freedom from the known*. Krishnamurti Foundation India, Chennai.

Noh, J. E. (2021). Review of human rights-based approaches to development: Empirical evidence from developing countries. *The International Journal of Human Rights, 26*(5), 883–901. https://doi.org/10.1080/13642987.2021.1981869

Sengupta, A. (2004). The human right to development. *Oxford Development Studies, 32*(2), 179–203. https://doi.org/10.1080/1360081041000169948

Stiglitz Joseph, E. (2024). *Road to freedom: Economics and society*. Random House.

Thurman Robert, A. F. (2012). *The Tibetan book of the dead: Liberation through understanding in the between*. Harper Collins Publishers.

Uvin, P. (2007). From the right to development to the rights-based approach: How 'human rights' entered development. *Development in Practice, 17*(4–5), 597–606. https://doi.org/10.1080/09614520701469617

Vandenhole, W., & Gready, P. (2014). Failures and successes of human rights-based approaches to development: Towards a change perspective. *Nordic Journal of Human Rights, 32*(4), 291–311. https://doi.org/10.1080/18918131.2015.957458

Xiang, S., Ping, Z.-B., Dong, Z.-F., Chen, K.-L., Zhu, X.-D., Larry Li, B., Tan, X.-Y., Zhu, B.-K., Liu, X., Zhou, C.-C., Fang, S., & Xiong, W. (2021). Resources and environmental costs of China's rapid economic growth: From the latest theoretic SEEA framework to modeling practice. *Journal of Cleaner Production, 315*. https://www.sciencedirect.com/science/article/pii/S0959652621023441

Yefeng, J., Tian, S., Xu, Z., Gao, L., Xiao, L., Chen, S., Xu, K., Chang, J., Luo, Z., & Shi, Z. (2022). Decoupling environmental impact from economic growth to achieve sustainable development goals in China. *Journal of Environmental Management, 312*. https://doi.org/10.1016/j.jenvman.2022.114978

Chapter 4
Alternative Economic Systems: Modern Versus Ancient

1 Context

Following the critical examination of the post-growth approaches in Chapter Three, the present chapter looks into some of the alternative economic systems that can provide insights and learnings that can help improve the present system. These include modern as well as ancient economic systems. In the process, the weaknesses of the modern systems are indicated and addressed by incorporating the good ideas from the ancient economic systems. Here, the idea is to think about an alternative system that would integrate the strengths of both approaches to create a more sustainable and equitable model. Throughout history, human beings have organized societies and economic systems to promote the wellbeing of people and the environment. These social and economic systems have evolved from ancient times and have taken a variety of shapes in contemporary periods. The ancient, especially the Indian economic system, is deeply rooted in philosophical and ethical principles, in contrast to the modern economic systems prioritizing efficiency and market mechanisms. Ancient Indian economics, as articulated in texts like the Arthashastra, emphasized moral values, social stratification, and the interconnectedness of social, political, and economic systems. This foundational philosophy continues to influence contemporary economic thought, particularly in areas such as sustainable development and social justice. Consequently, several new economic development paradigms have emerged to ensure the wellbeing of both people and the planet. However, as discussed in previous chapters, the prevailing modern economic systems have failed to tackle several contemporary social and economic challenges facing the world today. This chapter attempts to critically review and analyse both ancient and contemporary modern economic systems and suggests an alternative economic system that may resolve the current economic and social crises.

V. R. Reddy and B. Behera, *Understanding the Current Global Socio-economic and Ecological Crises*, https://doi.org/10.1007/978-3-031-98600-0_4

2 Modern Economic Systems

Modern economics, particularly the present, is grounded in classical and neoclassical economic theory. It focuses on market-based solutions, efficiency, and growth, often emphasizing the rational self-interest behaviour of individuals (assuming individuals act to maximize their utility or profits), market equilibrium (prices and quantities in markets are determined by supply and demand), economic growth (GDP growth is a key indicator of economic health and societal progress), scarcity and allocation (how to allocate scarce resources among competing uses), and efficiency (optimizing the use of resources to produce maximum output). Modern economics is concerned with maximizing material wealth, increasing efficiency, and fostering continuous economic growth through free markets, innovation, and competition.

Hence, the modern economic system is a complex and dynamic framework that governs how resources are produced, distributed, and consumed. It has evolved significantly and has been influenced by technological advancements, globalization, and policy decisions. Today, most economies operate under a mixed economic system, incorporating elements of capitalism and government intervention to balance efficiency and equity. In modern economies, market forces such as supply and demand determine the production and pricing of goods and services (Tian & Chen, 2022). However, governments intervene to regulate industries, provide public goods, and address economic inequalities. This balance ensures economic stability and promotes social welfare. Economic globalization has led to increased trade and financial integration across countries in the world today. Businesses and consumers benefit from access to a wide range of goods and services, while nations specialize in industries with a comparative advantage. However, globalization also presents challenges like economic dependency and market volatility.

As discussed in the previous chapter, while modern economic systems offer numerous benefits, they face significant challenges, including income inequality, economic recessions, environmental degradation, and geopolitical tensions (Lavie, 2022). Also, the previous chapter discussed the various strategies societies adopted to promote inclusive growth and address emerging global challenges. However, all the strategies seem to have failed as the global challenges are multiplying and remain unresolved. The weakness of the modern capitalist economic system has compelled societies across the globe to explore and innovate various alternative economic systems that can potentially replace the modern economic system. What follows is a brief review of some of the ancient and contemporary economic systems that societies have developed. Several alternative economic systems, both contemporary and ancient, challenge the dominant modern economic system, often represented by capitalism or neoliberal economics. These alternative economic systems tend to address inequality, environmental degradation, and social justice and overemphasize profit maximization.

2.1 Socialism

A socialist economic system is an economic model in which the state or the community owns and controls the means of production, distribution, and exchange (Newman, 2005). Unlike capitalism, which prioritizes private ownership and market-driven economies, socialism aims to reduce income inequality and ensure that resources are distributed more equitably. In socialism, the government typically owns and manages key industries such as healthcare, education, transportation, and energy. This prevents monopolies and ensures that essential services are accessible to all citizens. Unlike market economies that rely on supply and demand, socialist economies are often planned by the government. Economic planners decide what goods and services should be produced, in what quantities, and at what price, aiming to meet the needs of society rather than maximize profits.

One of the core objectives of socialism is to reduce economic inequality. Policies such as progressive taxation, free public services, and state welfare programs help distribute wealth more fairly, ensuring that basic needs like education, healthcare, and housing are met for all citizens. While some socialist systems allow small-scale private businesses, the state usually controls large industries and significant economic sectors. This prevents individuals from accumulating excessive profit and focuses on collective welfare. Socialist economies prioritize the well-being of citizens by providing social security, unemployment benefits, and pensions. This system reduces poverty and provides economic stability, ensuring people are not left behind due to market fluctuations.

While socialism promotes equality and social welfare, it also faces challenges. Centralized planning can sometimes lead to inefficiencies, resource misallocation, and lack of innovation due to limited competition (Allison, 2023). Additionally, excessive government control can reduce individual economic freedom and discourage entrepreneurship. While it offers significant benefits regarding social welfare and economic security, its effectiveness depends on how well it is implemented. Experience showed that many socialist countries experienced extreme poverty and environmental degradation (Allison, 2023). Hence, many modern economies adopt a mixed approach, incorporating socialist principles such as public services and social welfare within a broader market-driven framework. For example, countries like Cuba or the former Soviet Union practiced socialism, although modern democratic socialism (e.g., in Scandinavian countries) integrates elements of both socialism and capitalism. The socialist economic system encompasses various models prioritizing collective ownership and democratic management of resources. This system diverges from traditional capitalism by emphasizing social welfare and equitable distribution of wealth. Recent discussions highlight the evolution of socialist economies, particularly in countries like China and Vietnam, which have adopted a socialist market economy framework.

2.2 Solidarity Economics

Solidarity economic system (SES) is an alternative economic model prioritizing cooperation, social wellbeing, and collective decision-making over profit maximization. The history of the solidarity economic system reflects a significant evolution from traditional economic models toward cooperative and community-oriented practices. The history of solidarity economics includes African Americans establishing cooperatives in response to market failures and exclusion, creating alternative economic solutions for survival, democratic ownership, and profit-sharing, rooted in mutual aid and communalism, particularly during racial capitalism and economic discrimination (Gordon-Nembhard & Ifateyo, 2023).

The solidarity economic system emerged as a response to the limitations of capitalist economies, particularly during the nineteenth century, with the establishment of cooperatives and mutual aid societies. In recent decades, international organizations have recognized SES as vital for achieving Sustainable Development Goals (SDGs), promoting a shift towards regenerative economic systems (Egorov & Inshakov, 2023; Egorov et al., 2023). This system emphasizes mutual aid, cooperation, and ethical considerations in economic activities, marking a shift from profit-driven paradigms to those focused on social welfare and sustainability. Rooted in principles of social justice, mutual aid, and sustainability, this approach seeks to build an economy that serves the needs of communities rather than corporations. It challenges traditional capitalist structures by emphasizing solidarity, democracy, and ethical economic practices.

Unlike market economies emphasizing competition, solidarity economics promotes collaboration among individuals, businesses, and communities. Cooperatives, worker-owned enterprises, and non-profit organizations play a significant role in this system. In solidarity economics, economic relationships are based on cooperation rather than competition. People work together for the common good, supporting each other's needs through practices such as cooperatives, community-supported enterprises, and mutual aid networks. Decision-making in solidarity economics is inclusive and participatory. Workers, consumers, and local communities have a say in managing businesses and resources, ensuring that economic power is more evenly distributed. This model integrates ethical considerations into economic activities. It prioritizes fair wages, equitable working conditions, and sustainable practices to protect people and the environment. The idea is that everyone in a community can benefit from working collectively. Instead of focusing on individual success, solidarity economics fosters a sense of collective responsibility and shared prosperity.

Solidarity economics supports local economies by promoting small businesses, cooperatives, and fair-trade networks, which reduces economic dependency on large corporations and strengthens community resilience. Instead of measuring success through economic growth alone, solidarity economics considers human development, social inclusion, and environmental sustainability as key progress indicators. Workers own and manage businesses where profits and decision-making are shared collectively. Ethical trade practices that ensure producers receive fair compensation for their work. Local food systems that connect farmers directly with consumers, promoting sustainable agriculture.

Solidarity economics focuses on reducing inequality and promoting social and economic justice. It aims to address systemic inequalities by ensuring that marginalized and underrepresented groups have access to economic opportunities, resources, and decision-making power. Fair distribution of resources is a central goal, meaning wealth and opportunities are shared more equitably across society, especially among vulnerable populations.

One of the largest and most well-known examples of solidarity economics is the Mondragon Corporation in Spain's Basque Country. It is a federation of worker cooperatives operating in various industries, from manufacturing to retail. Workers are the owners and decision-makers, and profits are reinvested in the community. Mondragon Corporation, located in Spain's Basque Country, is a unique conglomerate of cooperatives that exemplifies the principles of cooperative economics. Comprising 81 cooperatives and numerous subsidiaries, it operates across various sectors, employing approximately 70,500 individuals and generating around 11 billion euros in revenue annually (Etxeberria-Otadui, 2024). The corporation's resilience and adaptability in economic challenges highlight its innovative governance and commitment to cooperative values.

Brazil has a strong solidarity economy movement, with thousands of cooperatives, social enterprises, and grassroots organizations (Dubois, 2021; Lemaître & Helmsing, 2012). The country has even established Solidarity Economy Forums and policies to support the growth of these initiatives, especially in marginalized communities. Brazil's solidarity economy movement encompasses various initiatives to promote social and economic equity through collective action. Notable examples include cooperatives formed by waste pickers, community banks, and women's solidarity networks, each contributing to local economies and social cohesion.

The Associação de Recicladores das Águas Lindas (ARAL) in Belém, Pará, exemplifies a solidarity economic enterprise where waste pickers organize democratically to improve working conditions and income generation (Cardoso et al., 2024). Members engage in self-managed activities, facing challenges transitioning from an association to a cooperative model, yet benefiting from better security and organization (Cardoso et al., 2024). Banco Palmas is another prominent example of a community bank that evolved into a solidarity fintech, providing microcredit and digital currency to support low-income populations, particularly during the COVID-19 pandemic (Diniz, 2023). This initiative has significantly contributed to local economic resilience by circulating money within communities (Diniz, 2023).

Time banking is another classic example of a solidarity economic system, which is a community-driven initiative that allows individuals to exchange services based on time rather than money, fostering social connections and mutual support. Various examples illustrate its implementation across different contexts, showcasing its adaptability and benefits. Time banks in the UK serve as an alternative social movement, challenging conventional economic systems by facilitating resource mobilization and social dynamics. They provide a platform for community members to exchange services, enhancing social cohesion and addressing local needs (Maharsi et al., 2023). Research on time banks in Barcelona indicates their effectiveness in generating social capital, focusing on bonding and bridging social capital among members. The study highlights the importance of reciprocity and intergenerational exchanges, which contribute to community resilience (Carnero et al. 2024).

While solidarity economics offers a more equitable and sustainable model, it faces numerous challenges, such as limited funding, lack of government support, and competition from dominant capitalist enterprises. However, with growing interest in social entrepreneurship and ethical business practices, solidarity economics is gaining recognition as a viable path toward a more just and inclusive economy. Solidarity economics provides an alternative economic development vision that prioritizes people and the planet over profits. Fostering cooperation, sustainability, and social justice presents a pathway toward a more democratic and humane economic system. As global challenges such as inequality and climate change intensify, solidarity economics offers innovative solutions for a fairer future.

While solidarity economics works well at the community level, critics argue that it can be challenging to scale these initiatives to address more significant national and global economic issues. The decentralized, localized nature of solidarity economics may not easily lend itself to large-scale industrial economies. Some solidarity economy enterprises rely on state subsidies or support, leading critics to question their long-term viability in a competitive market. Solidarity enterprises operating in a capitalist economy often face pressure to conform to market demands, which may undermine their values of cooperation and community welfare.

Many solidarity economy initiatives are criticized for their reliance on state support, which can undermine their autonomy and effectiveness. This dependency can lead to a dilution of the original goals of self-management and mutual aid, as these initiatives may become entangled in bureaucratic processes (Parnell, 2021). Critics argue that the concept of solidarity economy can distract from the development of authentic cooperative enterprises, which are essential for grassroots self-help (Parnell, 2021). The focus on solidarity can sometimes overshadow the need for practical, sustainable economic models that prioritize community resilience over ideological commitments (Lazányi, 2022). The theoretical frameworks surrounding solidarity economics are often seen as insufficiently rigorous, lacking the depth needed to address complex socio-economic realities. Critics advocate for a more integrated approach that combines ecological sustainability with social justice, emphasizing the need for continuous reflection and adaptation in practices (Lazányi, 2022).

2.3 Wellbeing Economics

The history of wellbeing economics reflects a shift from traditional economic indicators, such as GDP, towards a more holistic understanding of human and planetary welfare. It emphasizes that the ultimate goal of economic policy should be to enhance people's lives rather than simply increase economic output. This evolution has been influenced by various global initiatives and theoretical frameworks that emphasize sustainability and social equity. The New International Economic Order (NIEO) in 1974 aimed to create a fairer global economy but was undermined by neoliberalism in the 1980s, highlighting the challenges of implementing well-being-focused policies (Legge, 2025). The concept of a happiness economy has gained traction, prioritizing subjective well-being over material wealth (Agrawal et al., 2024).

The Wellbeing Economy Alliance (WEAll), established in 2018, promotes economic systems centred on wellbeing, with countries like New Zealand and Iceland leading the way in policy innovation (Hough-Stewart & Janoo, 2024). Finland's integration of well-being economics into its welfare model demonstrates a successful application of these principles, promoting inclusive growth and sustainability (Kyyrönen, 2024). Well-being economics draws from ecological and feminist economics, emphasizing the need for policies that respect ecological constraints and promote social justice (Hensher, 2024). This also primarily draws on principles from welfare economics, ecological economics, and social psychology, aiming to balance economic growth with social and environmental sustainability.

Traditional economics often prioritizes GDP growth as the leading indicator of a nation's progress. Wellbeing economics, however, argues that wellbeing encompasses more than just material wealth. It includes health, education, happiness, work-life balance, mental health, and environmental quality. Wellbeing economics asks: *Are people healthier, happier, and more fulfilled in their lives?* Rather than just *Is the economy growing?* While GDP measures the total market value of goods and services produced in an economy, it doesn't account for income inequality, environmental degradation, or the quality of life (Behera, 2023). Wellbeing economics advocates for the use of multidimensional indicators such as the Human Development Index (HDI), which measures health, education, and income; the Genuine Progress Indicator (GPI), which adjusts GDP by accounting for factors like income distribution, environmental costs, and social wellbeing, Happiness Index, which assesses national happiness levels, and Social Progress Index, which measures basic human needs, wellbeing foundations, and opportunity for individuals to improve their lives.

Well-being economics highlights that economic growth at the cost of environmental degradation is unsustainable in the long run. A healthy environment is foundational to human well-being, and policies should aim to protect ecosystems, reduce pollution, and combat climate change. It stresses the importance of 'sustainable development that meets the needs of the present without compromising the ability of future generations to meet their own needs' (WECD, 1987). Well-being economics emphasizes reducing inequality and ensuring everyone can access opportunities, resources, and social services. Economic policies should aim to reduce disparities in income, education, healthcare, and employment opportunities. It promotes social capital—the strength of communities, social bonds, and civic participation—is considered a key component of wellbeing. Unlike traditional economics, which focuses on labour productivity and employment rates, wellbeing economics also values the quality of work. It considers job satisfaction, work-life balance, and mental health as critical factors contributing to overall wellbeing. Policies encouraging shorter working hours, flexible work arrangements, and fair wages are promoted.

Wellbeing economics integrates insights from psychology, recognizing that mental health, stress levels, and overall happiness are crucial components of well-being. Governments should design policies to enhance people's mental and emotional well-being, such as investing in mental health services, promoting social connections, and reducing loneliness. Social cohesion, strong relationships, and a sense of community are essential for human wellbeing. Wellbeing economics argues that economic policies should focus on individual prosperity and fostering strong, supportive communities. Wellbeing economics places a high value on providing public goods such as education, healthcare, transportation, and green spaces. These services are fundamental to improving quality of life and should be accessible to all citizens, regardless of income. Governments play a critical role in promoting well-being through policy interventions. Wellbeing economics advocates for proactive government measures such as: Progressive taxation to reduce inequality, Social safety nets like unemployment benefits and healthcare, Investments in public health, education, and infrastructure that directly improve people's well-being.

Various countries and cities have begun implementing frameworks and policies reflecting practical examples of wellbeing economics. New Zealand is one of the leading nations in implementing wellbeing economics. In 2019, the country introduced the Wellbeing Budget, prioritizing mental health, child poverty, inequality, and environmental sustainability over GDP growth alone. The budget assesses policies based on their contribution to the population's wellbeing rather than just economic outputs (Hayden & Dasilva, 2022). New Zealand's Living Standards Framework incorporates specific measures for different population subgroups and Wales' Well-being of Future Generations Act, mandating government alignment with wellbeing goals across all departments and public bodies (Robinson, 2023).

Bhutan's Gross National Happiness Index is one of the earliest examples of well-being economics. GNH measures prosperity not through GDP but through spiritual, cultural, and ecological health. It includes indicators like psychological well-being, education, health, living standards, and ecological diversity (Chetri, 2023). The Gross National Happiness (GNH) Index represents a holistic approach to measuring the wellbeing of a nation beyond traditional economic indicators. It emphasizes the importance of psychological, physical, societal, and environmental wellbeing, aiming to create conditions for human flourishing. The GNH Index is structured around four pillars and nine domains, utilizing 33 indicators to assess collective happiness and guide policy decisions (Chetri, 2023). The Four Pillars are sustainable development, cultural preservation, environmental conservation, and good governance. The Nine domains include psychological wellbeing, health, education, time use, cultural diversity, good governance, community vitality, ecological diversity, and living standards. Indicators include measures of social connections, spirituality, and time use, providing a comprehensive view of happiness (Chetri, 2023).

GNH has inspired other nations to consider wellbeing in their development policies, promoting a shift from GDP-centric models to those prioritizing happiness and sustainability (Dash, 2024; Ongsod & Braila, 2022). Bhutan's GNH model is a case study for integrating qualitative measures into governance, potentially enhancing citizen engagement and satisfaction globally (Ongsod & Braila, 2022). While GNH offers a promising alternative to traditional economic measures, critics argue that it may not fully capture individual happiness, as it focuses on societal conditions rather than personal experiences (Chetri, 2023).

Scotland's National Performance Framework (NPF) is a comprehensive tool for measuring the country's progress towards sustainable and inclusive growth. Established to align government activities with 11 National Outcomes, the NPF emphasizes wellbeing and aims to create a unified approach across public services. This framework assesses economic performance and integrates social and educational dimensions, reflecting a holistic view of progress in Scotland. The NPF includes economic indicators such as GDP growth, employment rates, and income inequality, which benchmark Scotland against other OECD countries (McDonald & Richmond, 2019). It promotes inclusive growth by focusing on exports, innovation, and fair work practices, aiming to create a balanced economic landscape (McDonald & Richmond, 2019).

The NPF has reoriented government priorities towards wellbeing, fostering collaboration among public bodies and enhancing early intervention strategies (Wallace, 2019). It has also reframed the relationship between the Scottish Government and local authorities, although challenges remain in citizen engagement (Wallace, 2019). While the NPF has made significant strides in promoting inclusive growth and wellbeing, critiques highlight the need for deeper citizen engagement and more effective participatory measures to realize its potential (Wallace, 2019) fully. Wellbeing economics redefines economic success, arguing that economic policy's ultimate goal should be to improve the quality of life for all people, not just to increase economic output. It shifts the focus from material wealth to holistic well-being, advocating for socially inclusive, environmentally sustainable policies that are aligned with human happiness and health. This approach challenges the traditional economic paradigm by integrating ethics, social justice, and environmental sustainability into economic decision-making.

Critics of wellbeing economics highlight several fundamental issues regarding its assumptions, methodologies, and implications for policy-making. These critiques emphasize the need for a more nuanced understanding of wellbeing that transcends simplistic measures of happiness and utility. Critics argue that wellbeing economics often relies on self-reported happiness, which may not accurately reflect true societal welfare. This reliance on subjective measures overlooks the evolutionary basis of emotions, suggesting that such metrics are inadequate for normative economic analysis (Burnham & Phelan, 2023). The assumption that happiness can be quantified and used as a basis for policy decisions is contested, as it may lead to technocratic governance that neglects individual and community contexts (Singh & Alexandrova, 2020). Welfare economics is critiqued for ignoring the institutional frameworks that shape market outcomes. The political and social contexts in which economic transactions occur are essential for understanding their impact on welfare (Leschke, 2016). A more comprehensive approach to wellbeing should consider moral and ethical dimensions, as well as the capabilities and functioning of individuals within their social environments (Choudhury, 2020).

The modern economic systems that are currently prevailing in the world largely focuses on market-driven principles, often side-lining ethical considerations in favour of efficiency (Akolkar et al., 2024), which faces challenges from globalization and technological advancements, which require adaptive policies that may conflict with traditional values (Gupta, 2024). There is a growing interest in integrating ancient principles into modern policy-making to address issues like poverty and sustainable development (Gupta, 2024). What follows is a detailed review of some of the prominent ancient Indian economic thoughts such as *Arthasastra*, Buddhist economics and Gandhian (Hindu) economics.

3 Ancient Indian Economic Principles

Ancient Indian economics integrated ethical considerations, emphasizing karma and Dharma in economic activities (Malov & Popov, 2024). The Varna and Ashrama systems structured labour and economic roles, promoting a sense of duty and social responsibility (Shukla, 2024). Kautilya's Arthashastra text provided insights into taxation, trade, and resource management, advocating for a balanced approach to governance and economic prosperity (Gupta, 2024).

3.1 Arthashastra

The *Arthashastra*, written by Chanakya (also known as Kautilya or Vishnugupta) in ancient India around the fourth century BCE, is a foundational text on statecraft, economics, military strategy, and governance. Chanakya was the chief advisor to Chandragupta Maurya, the founder of the Maurya Empire, and his *Arthashastra* provides an in-depth analysis of how to govern an empire efficiently. The *Arthashastra* primarily treats governance and politics (Patil, 2024). Chanakya emphasizes the importance of a strong and centralized state, where the ruler is responsible for maintaining order and ensuring prosperity. The ruler is advised to act pragmatically, placing the state's welfare above personal interests. The king is seen as the central figure in governance, but his rule must be just and ethical to ensure stability. He must protect his people and maintain internal and external security. Chanakya advises the ruler to rely on a council of wise ministers for advice on state matters, emphasizing collective decision-making. A strict legal system is necessary to maintain order. Chanakya outlines various punishments and laws to ensure justice and deter crime.

The *Arthashastra*, authored by Kautilya (Chanakya), presents a comprehensive framework for understanding economics in ancient India, emphasizing the interplay between governance and economic policies. Its principles remain relevant today, offering insights into taxation, trade, and resource management. The following sections elaborate on key economic doctrines from the *Arthashastra*.

The *Arthashastra* is known for its detailed analysis of economic policies and the administration of the state's finances. Chanakya understood the importance of a strong economy in maintaining political power. The text emphasizes efficient tax collection and the need for a well-maintained treasury. Taxes should be fair but necessary to fund the army, administration, and public projects. Farmers were expected to pay the state a portion of their produce as tax (Priya, 2024). Trade was encouraged, and the state was responsible for regulating markets, setting prices, and controlling weights and measures (Gupta, 2024). Chanakya encouraged state investment in infrastructure like roads, irrigation, and agriculture to ensure long-term economic prosperity. Chanakya believed in a well-regulated economy where the state had a key role. He promoted state monopolies in essential industries like mining, salt production, and arms manufacturing to ensure state control over critical resources.

Chanakya's *Arthashastra* emphasizes the welfare of the people as a crucial duty of the state. It discusses how a prosperous population is the foundation of a stable and powerful state. The king is responsible for ensuring the welfare of his subjects. This includes providing relief during famines, building public facilities (such as temples, roads, and irrigation systems), and regulating wages. Given that agriculture was the backbone of the economy, special attention was given to the protection of farmers and promoting agricultural productivity. One of the defining characteristics of the *Arthashastra* is its pragmatic and often ruthless approach to statecraft. While Chanakya acknowledges the importance of ethics and Dharma, he believes that the survival and strength of the state take precedence. This realism can be seen in his governance, diplomacy, and war advice. Chanakya's approach has often been compared to Machiavelli's *The Prince* due to its pragmatic, sometimes ruthless approach to ruling. For instance, he suggests using spies, deception, and even assassination to maintain power if necessary. Chanakya's ethics can be seen as utilitarian. He argues that the end (the state's welfare) justifies the means, and the ruler should take whatever steps are necessary to secure the state, even if they involve morally questionable actions.

The *Arthashastra* provides a comprehensive framework for understanding investment and trade in ancient India, emphasizing the state's role in regulating economic activities. Kautilya viewed trade as a vital component of the economy, advocating for private entrepreneurship under strict state oversight to ensure revenue generation for the monarchy. This foundational text outlines various economic principles that remain relevant today. Kautilya identified trade as a primary source of revenue, alongside agriculture and animal husbandry (Das & Mahapatra, 2012). The state was responsible for regulating trade practices to prevent fraud and ensure fair practices, thus protecting both traders and consumers (Bhat, 2023). The Arthashastra emphasizes the importance of a centralized administration for effective revenue collection from trade activities. Kautilya advocated for investments in infrastructure to support trade, such as roads and markets, enhancing economic connectivity (Manrai & Goel, 2017). The text also discusses the significance of treasury management and prudent expenditure allocation to foster a conducive environment for trade (Bhat, 2023). While Kautilya's principles focus on state control and regulation, some modern perspectives argue for a more liberalized approach to trade, emphasizing the benefits of free markets and reduced government intervention. This contrast highlights the ongoing debate about balancing state oversight and market freedom in economic governance. Chanakya's *Arthashastra* remains a landmark text in ancient Indian political thought, combining ethical principles with a pragmatic approach to governance, economics, and strategy, reflecting a holistic vision for running a prosperous and powerful state.

3.2 Buddhist Economics

Buddhist economics represents a unique economic theory and practice approach, emphasizing ethical considerations, sustainability, and social welfare. Rooted in Buddhist teachings, it advocates for a balance between material and spiritual wellbeing, contrasting sharply with capitalist models prioritizing wealth accumulation. This perspective is increasingly relevant in addressing contemporary economic challenges. Buddhist economics is grounded in ethical principles derived from the *Suttapiṭaka*, promoting honest livelihoods and non-exploitation (Bawono et al., 2023). It emphasizes moderation, non-violence, and generosity to reduce suffering and foster harmony within communities (Amore, 2024). The framework advocates for sustainable practices that protect the environment while addressing inequality, as highlighted by Clair Brown's vision for a caring economy (Brown, 2023). Buddhist economics offers a holistic alternative to neoliberal capitalism, focusing on shared prosperity and wellbeing rather than mere income growth. It encourages policies that align with altruistic behaviours, recognizing the interconnectedness of societal health and environmental sustainability (Brown, 2023).

Schumacher (1973) was the first to synthesise Buddhist economic thoughts, and he outlined the virtue of Buddhist economics for a sustainable and fulfilling economy. He argued that the goal of humanity should be spiritual liberation or freedom rather than everlasting bondage to physical or material things and that Buddhist economics offers a means to this end. In many ways, Buddhist economics is fundamentally different from the modern economic system. For example, modern economics is primarily concerned with the production of economic goods. In contrast, Buddhist economics is only interested in human liberation/freedom from bondage, and the social and economic life of people should be organised in such a manner that helps achieve this objective. According to the Buddhist way of economic life, the consumption of goods should be satisfying rather than saturating (Rosser & Rosser, 2018). Modern economists treat human labour or work as a necessary evil that needs to be reduced to a minimum. For instance, for an employer, it is merely an item of costs of production which needs to be minimised, and for employees, it is disutility (as it involves the sacrifice of one's leisure and comfort). In the case of Buddhist economics, to work is to purify one's character by seeing oneself clearly in the mirror of a relationship with others but not to multiply wants/desires. Buddhist teachings and philosophy clearly distinguish between physical and psychological aspects of wealth. According to them, having wealth does not stand in the way of liberation or freedom, but the attachment to wealth does; similarly, it is not the enjoyment of pleasurable things that create an obstacle for achieving freedom, but the craving for them does. In the Buddhist way of economic life, having the right livelihood is of utmost importance for treading on the path of liberation, and that is why Right Livelihood is prominently placed in the Noble Eightfold Path of Buddha's teaching (Schumacher, 1973).

One of the most striking messages of Buddhist economics is the maintenance of harmony between human society and nature. Buddhists stress that plants and animals should be preserved for long-term ecological balance. Non-renewable resources should not be wasted as they cannot be renewed once they are exhausted. Local production should be favoured over long-distance trade, in order, to reduce wasteful transportation activities and to encourage self-sufficiency (Schumacher, 1973). According to Buddhist teachings, every ecosystem provides sufficient material for the sustenance of all living beings. Hence, simple and non-violence living brings harmony between humanity and nature.

While Buddhist economics presents a compelling framework for addressing modern economic issues, critics may argue that its implementation faces challenges in a predominantly capitalist world, where profit motives often overshadow ethical considerations. Nonetheless, the integration of Buddhist principles into economic practices continues to inspire movements toward more equitable and sustainable systems.

3.3 Hindu Economics

Hindu economics, often called *Sanatan* Economics, is rooted in the ethical and moral principles derived from Hindu scriptures such as Vedas, Upanishads, Sutras, etc. It emphasizes sustainability, self-sufficiency, and balanced economic growth, aligning economic activities with societal wellbeing and moral values. This framework has evolved alongside historical economic practices in India, particularly during the colonial period and the subsequent liberalization in the 1990s, reflecting a complex interplay between traditional Hindu values and modern capitalist dynamics. Central to Hindu economics is Dharma, which promotes ethical actions in harmony with universal laws (Vinod & Deodhar, 2024). Sanatan Economics advocates for responsible resource use and value-based economic decisions, aiming for societal betterment (Jakhar et al., 2023). The following are the pillars on which the principles of Hindu economics are based.

In Hindu economics, Dharma plays a central role. Dharma refers to righteous behaviour and moral duties that each individual must follow in their economic and social activities. Economic actions must align with the principles of fairness, justice, and moral responsibility (Vinod & Deodhar, 2024). Economic activity is seen as a form of duty, not just a means to accumulate wealth. Wealth should be earned in accordance with Dharma, meaning through ethical and non-exploitative means (Sihag, 2005).

Artha, one of the four *purusharthas* (goals of human life), refers to the pursuit of wealth and material prosperity. In Hindu economics, the pursuit of wealth is accepted, but it must be balanced with Dharma and other spiritual goals. Excessive attachment to wealth is discouraged and should be used for the greater good. Wealth is necessary to live a balanced life and fulfil social and familial duties, but greed is condemned. Wealth should be used to support oneself, family, and society while also contributing to charitable and social causes (Vinod, 2012).

Kama, another of the *purusharthas*, refers to the pursuit of pleasure and desires. In Hindu economics, pursuing pleasure is natural, but it should be within the bounds of self-control and not at the cost of moral principles or societal wellbeing. Economic activities can fulfil desires but should not be indulgent or selfish. There should be a balance between material pleasures and spiritual wellbeing.

Moksha represents the ultimate goal of life in Hindu philosophy: liberation from the cycle of birth and death (samsara). In economic terms, Hindu economics sees wealth accumulation as a means to an end, not the end itself. Wealth should support spiritual growth, not detract from it. Economic success is seen as transient, and the focus should remain on inner growth and freedom from attachment to material things.

Charity or *dana* is an essential part of Hindu economics. The ancient texts advocate for sharing wealth with those in need and emphasize that wealth should be used for the welfare of society, particularly for the upliftment of the poor and disadvantaged. *Dana* should be done selflessly, without expecting anything in return. Wealth is seen as a form of stewardship, where the rich must support the less fortunate.

Hindu philosophy deeply reveres nature, often personified as *Prakriti* (nature) or *Bhoomi* (earth). Hindu economics advocates for the sustainable use of natural resources, avoiding over-exploitation and maintaining harmony with the environment. Economic activity should respect and protect natural resources, as nature is sacred. Excessive consumption and moderation are discouraged in personal and economic life.

In traditional Hindu society, economic roles were often linked to the *varna* system (social classification), which divided people into four groups: Brahmins (priests and scholars), Kshatriyas (warriors and rulers), Vaishyas (merchants and farmers), and Shudras (laborers). While this system is controversial in modern times, the underlying principle was that different people had different economic and social responsibilities, and fulfilling one's economic duty (according to one's talents and abilities) was part of living a righteous life. In a way, it is very close to the division of labour proposed by Adam Smith in his Wealth of Nations in 1776. Economic activities were traditionally seen as a collective responsibility, with different groups contributing to society in specific ways.

Similar to Buddhist and Gandhian economics, Hindu economics encourages a life of simplicity and detachment from excessive materialism. The idea is to enjoy material prosperity without becoming attached to it, recognizing that material wealth is impermanent. The Bhagavad Gita teaches the importance of acting without attachment to the results, which also applies to economic activities. Wealth should not become an obsession, and the pursuit of spiritual wellbeing should always remain paramount. Hindu economics is a moral and spiritual framework for economic activity, encouraging balance between material prosperity, social justice, environmental stewardship, and spiritual fulfilment.

3.4 Gandhian Economics

Gandhian, often referred to as Hindu economic thought, is another alternative economic perspective similar to Buddhist economic thought (Kumarappa, 1951). Mohandas (Mahatma) Gandhi, the father of Indian independence, laid the foundation of Gandhian economic principles. He outlined the economic thoughts in his 1909 book Hind Swaraj. Swaraj means self-rule and self-sufficiency, both nationally and at the regional and local village level (Rosser & Rosser, 2018). Most of Gandhi's ideas about economic and social life are largely derived from the ancient Hindu scriptures, such as the *Upanishads*, the *Vedas*, and the Bhagavad Gita. Hence, it is often termed as Hindu economics.

The core idea of Gandhian economics lies in life based on truth, non-violence and non-possession or giving up (Kumarappa, 1951). It essentially rejects the concept of maximisation of self-interest, which is the main principle on which modern economic thoughts are based. Gandhi exalts that the earth provides enough material goods to satisfy every man's need but is insufficient to fulfil every man's greed. Gandhi's vision of the economy is not based on a parasitic economy where one agent lives on the other's effort, but it is based on a service or caring economy where resources are shared among all members of the economy based on their need. Like the honeybees who collect nectar from different flowers under a queen bee, gather in one place and share among themselves.

Gandhi was against an economic system that promotes a luxurious lifestyle disregarding the basic primary needs of people. Ethics, morality, and non-violence should be the guiding principles of any economy rather than only limited economic consideration. He cites the example of crop choices to be made for a plot of land, where there are two options available for a farmer. One option is to grow food grains which will nourish the hungry and promote life, and the other option is to cultivate tobacco which will fetch high monetary return, but it will take people's life; hence it is not based on non-violence and truth.

Although these two and other alternative approaches have been widely discussed in an academic circle in recent years, they are largely confined to intellectual deliberations and did not translate into policy levels and hence have very little impact on breaking the consumer-producer nexus. Why have we not applied these approaches? What have we missed?

3.5 Dharmic Capitalism

Balasubramanian (2022) introduced the concept of Dharmic Capitalism, a system where economic activities are intertwined with ethical and moral responsibilities. This model emphasizes the balance between wealth creation and societal welfare, suggesting that economic policies should be rooted in righteousness (*Dharma*). Dharmic Capitalism is an economic framework that combines principles of capitalism with values rooted in Dharmic philosophies, especially from Hinduism, Buddhism, Jainism, and Sikhism (Box 1). Unlike traditional capitalism, which primarily emphasizes profit maximization and market efficiency, Dharmic Capitalism advocates for profit in harmony with ethical responsibility, social welfare, and ecological balance. It prioritizes purposeful, sustainable, and compassionate business practices aligned with moral and spiritual obligations. Key aspects of Dharmic Capitalism include Buddhist Economics (focuses on compassion and interdependence, advocating for economic systems that promote wellbeing and sustainability (Gao, 2022). Conscious Capitalism: Incorporates ethical principles from Buddhism, aiming for a market that serves all sentient beings, particularly the impoverished (Boyle, 2022). Dharma in Business: Encourages leaders to make decisions that align with ethical values and the greater good, fostering a culture of mindfulness in corporate environments (Gor, 2010). Inner Truth: Emphasizes the importance of aligning personal and organizational values with Dharma, promoting integrity and purpose in business practices (Hawley, 1995). The following are the key tenets of Dharmic capitalism.

Dharmic Capitalism is built upon *Dharma*, which emphasizes righteousness, duty, and moral principles. Rather than seeking profit alone, it advocates for businesses to operate in a way that upholds ethical duties to society, the environment, and future generations. Dharma encourages firms to integrate ethical considerations into decision-making, balancing economic objectives with social and environmental responsibilities. Dharmic Capitalism promotes wealth generation that benefits all stakeholders-employees, communities, consumers, and the natural world. It recognizes that wealth should be used for personal gain and the welfare of society at large. This encourages inclusive business practices, fostering equitable distribution of wealth while ensuring long-term sustainability.

A fundamental *Dharmic* principle, *ahimsa*, or non-violence, extends to business activities by discouraging practices that harm the environment, exploit workers, or create social inequities. Dharmic Capitalism encourages organizations to adopt environmentally friendly practices, support fair labour conditions, and engage in ethical sourcing. In *Dharmic* Capitalism, businesses are encouraged to look beyond profit as their sole purpose. Instead, companies are seen as entities with the potential for positive societal impact. This means improving quality of life, contributing to community wellbeing, and solving pressing social or environmental challenges. This philosophy aligns with the idea that true wealth includes financial gain, social capital, and community welfare.

Box 1: Dharmanomics Dharmanomics as an economic system deeply rooted in moral and ethical considerations. Unlike purely profit-driven capitalist models or rigid state-controlled economies, this model integrates community welfare, sustainability, and economic justice. It argues that indigenous systems in India and Southeast Asia operated on principles that ensured economic growth without sacrificing social harmony. A significant aspect of Dharmanomics is the role of temples Srenis (guilds) as religious institutions and economic hubs. Examples from the Chola, Pallava, and Vijayanagara empires illustrate how temple economies functioned as self-sustaining economic ecosystems. Srenis are akin to modern-day corporations that played a crucial role in self-regulation, ensuring fair wages, protecting workers' rights, and promoting innovation. These ancient guilds like contemporary business models believed decentralized economic governance as a key to sustainable development. In fact, concepts like division of labour and free markets—existed in Indian economic systems centuries earlier. Indian economic models influenced civilizations in Cambodia, Vietnam, Thailand, and Indonesia that shaped their trade policies and governance. Principles like sustainability, ethical business practices, and decentralized governance is particularly are relevant even in today's world, where wealth inequality and environmental concerns dominate global discourse. Dharmanomics challenges conventional economic thinking and offers an alternative rooted in indigenous traditions with more ethical and sustainable financial system by adopting India's ancient economic wisdom.

Source Balasubramanian (2024).

Dharmic teachings emphasize a harmonious relationship with nature, recognizing humanity's interdependence with the environment. Dharmic Capitalism supports sustainable use of resources, minimizing waste and pollution, and prioritizing renewable energy sources. This stewardship reflects a commitment to ecological responsibility, protecting the environment for future generations. Leaders in Dharmic Capitalism are encouraged to embrace self-awareness, compassion, and humility, seeing their role as profit-makers and custodians of society's wellbeing. Spiritual development, mindful decision-making, and ethical behaviour are encouraged, fostering a culture of integrity and empathy within organizations.

Companies that adhere to Dharmic Capitalism often adopt policies that support their employees' wellbeing, practice fair trade, engage in philanthropy, and commit to environmental sustainability. Examples include businesses prioritizing employee welfare, creating socially impactful products, or investing profits into community projects. This approach aligns closely with modern movements like corporate social responsibility (CSR) and environmental, social, and governance (ESG) criteria, but Dharmic Capitalism differs in its foundational ethos. Where CSR and ESG practices may sometimes be add-ons to a fundamentally profit-driven model, Dharmic Capitalism considers ethical action, sustainable development, and societal welfare as its core, making profit a means to these ends rather than the end itself.

One challenge is balancing the demands of a competitive market with Dharmic values, which can sometimes increase operational costs or reduce profitability in the short term. However, advocates argue Dharmic Capitalism fosters long-term resilience, customer loyalty, and community support. As global interest grows in sustainable and ethically conscious business practices, Dharmic Capitalism presents an alternative path to economic development where profit, people, and the planet can thrive in harmony.

4 Human Behaviour and Economics: Inward and Outward Crises

As highlighted at the beginning of this chapter, the dominant economic framework across nations is the neoliberal modern economic system. This system, as previously discussed, prioritizes consumption over sufficiency, production over sustainability, economic growth over equitable distribution, and the accumulation of goods over their actual utility (Behera, 2012, 2017, 2023). Such priorities have entrenched consumerism and commercialism in most societies, precipitating ecological, socioeconomic, and political crises. This indicates that humanity is ensnared in a relentless cycle of satisfying endless desires and pleasures, which drive overproduction and pose serious threats to societal wellbeing and the planet's ecosystems.

The crises we face should be understood as both outward (ecological, economic, and social) and as reflections of inward disturbances within individual human beings (Behera, 2017). The disorder within human consciousness fuels external crises, and a comprehensive transformation across all spheres of human activity is necessary to create a new economic and social order. For instance, the deeply ingrained acquisitive tendency within individuals fosters societal inequality. The perpetual psychological need for security—manifested in the pursuit of status, wealth, power, and recognition—ironically undermines the physical security (access to food, shelter, and clothing) of the broader population. A crucial aspect often missing from academic and policy discussions is the recognition of the urgent need for a radical transformation of the human psyche, which lies at the root of most global problems. Critical questions arise: how can a deep, fundamental, and radical change in the human mind be realized? What sustains the disorder within the human mind? In other words, what inward crises have given rise to the outward crises we see today?

5 Physical Versus Psychological Wellbeing

One prominent disorder in the human mind is the failure to grasp the distinction between physical and psychological wellbeing. Contemporary societies largely focus on fulfilling individuals' material needs—such as food, shelter, clothing, and other basic amenities—while assuming that psychological happiness stems from material access. Yet, it can be argued that psychological wellbeing ultimately shapes an individual's material needs (Behera, 2017). As previously noted, the quest for psychological security impedes the universal provision of physical security. When individuals are inwardly impoverished—psychologically and spiritually—they attempt to enrich themselves through possessions, leading to increasingly complex demands and escalating societal issues (Krishnamurti, 2010a).

For instance, owning a house meets a fundamental physical need by offering protection against natural forces and societal threats. However, this need transforms into a psychological craving when pleasure is derived not from the use of the house, but from the mere ownership of it—the pleasure of possession. Similarly, individuals often seek happiness in the ownership of property, family, and adherence to various ideas and ideologies. Yet, the question remains: can true happiness be found in impermanent entities such as possessions, relationships, or political and religious ideologies? (Krishnamurti, 1947: 105). As long as individuals pursue lasting happiness through transient means, unhappiness will persist. True and lasting happiness arises only when the mind is liberated from attachments and psychological entanglements, achieving a "choiceless" state free from inner conflict (Krishnamurti, 2010a).

6 The Constant Pursuit and Fulfillment of Desire and Pleasure

Modern economics operates on the assumption that human wants and desires are limitless, positioning the satisfaction of these desires as a central objective of economic systems and governance. Consumer desire, in turn, drives the expansion of goods production, as producers continually offer a wider range of goods to satisfy an ever-growing array of consumer preferences. However, it is crucial to differentiate between human "needs" and "wants/desires" (Behera, 2012). Traditional economic analysis has predominantly focused on the fulfillment of desires, which are generated by the five senses and are virtually limitless.

In contrast, needs are essential for sustaining life and health. The core issue lies in humanity's inability to stop at satisfying basic needs; instead, individuals continue acquiring goods to fulfill psychological desires fueled by greed. Since desires originate from the interaction of sensory organs with mental feelings, they drive individuals to seek out material goods, and the satisfaction of such desires is endless, causing profound environmental and social harm (Jena & Behera, 2017). The compulsion to possess and consume persists even when it is destructive.

As previously discussed, humanity's insatiable desire to accumulate and consume is propelling both economic growth and environmental degradation. Psychological wellbeing is thus destabilized by the constant conflict of competing desires and the endless pursuit of pleasure (Behera, 2017). Figure 1 schematically illustrates the anatomy of human desire and the process of its cessation. The five senses—skin, eyes,

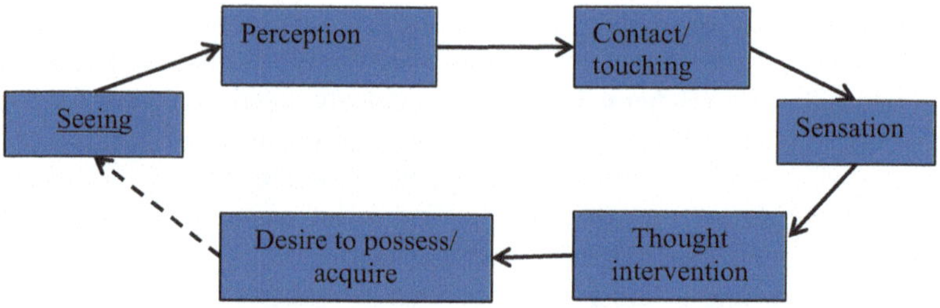

Fig. 1 Anatomy of human desire as described by Krishnamurti in his 1964 public talk (1964: 16). *Source* Authors

nose, ears, and tongue—merely respond to external stimuli; they do not inherently produce desires. Desire arises only when thought and feeling intervene in sensory experiences (Krishnamurti, 1964). Therefore, a crucial question emerges: can the senses operate independently, without the interference of thought?

A fundamental assumption underlying much of economic theory is the notion of the "rational" economic agent—an individual whose actions are governed by the singular aim of maximizing self-interest (Perman et al., 2003). In this view, self-centeredness becomes not only acceptable but essential to rational behavior. Yet one must ask: can such self-serving conduct truly be called rational? It is precisely this persistent selfishness, evident in the motives and daily actions of individuals, that lies at the heart of the profound social, economic, and ecological crises facing humanity today. The relentless pursuit of personal gain is also the root from which human corruption springs.

Emerging studies in the field of happiness economics reveal a deep paradox: that selfishness, rather than leading to satisfaction, often cultivates unhappiness (Lane, 2017). Thus, the pathway toward healing our fractured world may require nothing less than the dissolution of the individual ego—the "self" that clings so tightly to its desires and fears. This invites deeper questions: Is it possible for a human being to act selflessly, even amidst the demands of everyday life? Can one ever be free from the pull of self-interest? Why does selfishness arise at all? And ultimately, what is this "self" that governs so much of human behavior?

Figure 2 sketches the anatomy of the self—the ego or "me"—through which all action is filtered. The self is not innate but rather a construction, shaped by accumulated experiences, memories, and knowledge acquired throughout life, and sustained through the activity of thought (Krishnamurti, 2012). These fragments of past sensation and feeling, arising from the senses, are seized upon by thought and given continuity as the "me". The vital question, then, is whether human beings can live and act from a deeper intelligence—not one conditioned by past memories and experiences, but one free from the narrow bindings of thought. Can life be lived from a place untouched by the divisions of self, in pure awareness and compassion?

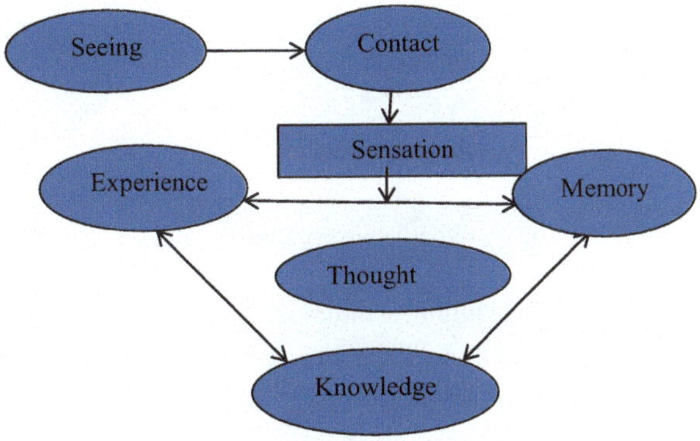

Fig. 2 The structure of the 'self', 'ego' or 'me'. *Source* Modified from Rodrigues (2001)

6.1 Sensory Value Versus Eternal Value

The modern economic system is fundamentally driven by benefit–cost analyses, where economic and social decisions are based on quantifiable measures of utility. Societies, by and large, assign value to the things of the Earth through the lens of the five senses—sight, sound, taste, touch, and smell—while largely neglecting the deeper, eternal value inherent in nature. The crises confronting the world today stem from this overemphasis on sensory value and the near-complete disregard for nature's timeless significance. As Krishnamurti (1945: 9) insightfully observed, "Efficiency becomes ruthless when sensory values predominate and the eternal value is set aside."

In modern, consumption-driven societies, nature has been approached through a utilitarian lens. Ownership, commodification, and pricing of natural resources have become standard practices, intended to facilitate the efficient use of the environment's assets (Perman et al., 2003; Hussen, 2004). Yet, vast segments of the natural world remain outside this system of valuation, giving rise to widespread disruptions and externalities that affect all of society—climate change being one of the most profound examples.

This utilitarian attitude has devastated the Earth; it reflects not love for the planet, but merely the exploitation of it. Humanity's relationship with nature has been reduced to one of use and control, rather than reverence and care. Thus, it is imperative that humans rediscover and cultivate a deep, intrinsic connection with the Earth—an understanding that transcends mere sensory satisfaction. As Krishnamurti (1975: 146) eloquently reminds us:

> If you lose touch with nature, you lose touch with humanity. If there's no relationship with nature then you become a killer; then you kill baby seals, whales, dolphins, and man either for gain, for sport, for food, or for knowledge. Then nature is frightened of you, withdrawing its beauty. You may take long walks in the woods or camp in lovely places, but you are a killer and so lose their friendship. You probably are not related to anything to your wife or your husband.

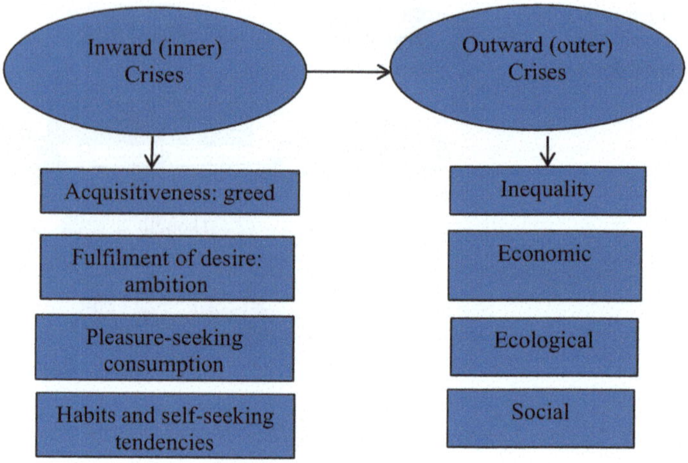

Fig. 3 The relationship between inward and outward crisis. *Source Authors*

The disorder within human consciousness, and its reflection in the outward crises of the world, may be more clearly understood through the schematic representation provided below (Fig. 3). Without inward order within each individual, outward harmony in society remains impossible. Thus, any true transformation of society—and of its economic structures—must begin with a radical renewal of human consciousness. The following section turns to explore the possible ways by which such an inner revolution might take place.

7 Transformation of the Human Mind and New Economy for Happiness

The preceding discussion on post-growth approaches highlights that while these models offer important insights, they remain largely idealistic and impractical when it comes to achieving sustainable and equitable development. This is primarily because the existing socio-economic structures and prevailing human behaviours are deeply rooted in the fundamental economic principles of selfishness, acquisitiveness, and competitiveness. These traits, forming the core of individual economic behaviour, propel humanity toward an endless pursuit of 'more,' as the marginal utility of wealth and money accumulation does not diminish.

It is argued that unless such desires are moderated, accumulation—and the exploitation it entails—will persist, whether through labour, as outlined in Marxist thought, or through resource depletion, as seen in environmental degradation paradigms. In response to the dangers of unchecked greed, various philosophical perspectives have emerged. Among the most prominent are Gandhian philosophy, Buddhist economics, the "Small is Beautiful" movement, and Deep Ecology. Modern philosophers such as Jiddu Krishnamurti have also strongly critiqued the destructive effects of excessive human desire. Significantly, many of these perspectives trace their roots back to ancient Indian wisdom, found in texts like the Vedas and the Epics. Drawing inspiration from these traditions, contemporary frameworks integrating social and ecological justice are being developed. However, the extent to which these philosophical underpinnings can catalyse meaningful behavioural change remains an open question.

For example, the eco-spiritual worldview, as described by Marale (2012), emphasizes love, compassion, care, and deep concern for the wellbeing of all beings in the universe. It proposes a holistic transformation of human behaviour and habits toward greater ecological sustainability. Central to this approach is the transformation of the human mind itself, recognizing that the mind shapes the world it perceives and creates. Ancient Indian scriptures also stress the importance of understanding the workings of the human mind, which in turn directly affects the natural and social environments. These traditions portray numerous instances where elements of flora and fauna are revered as divine, a practice still alive in certain communities in India that protect trees and animals with great devotion, prohibiting their harm.

Within this worldview, humans are regarded as integrated beings—possessing physical, intellectual, emotional, and spiritual needs—that must be fulfilled through righteous means. Crucially, it posits that there is no real division between the internal and external environments, as the external world is a projection of the mind. If individuals confine their desires to only what is essential, the mind can be brought under control; for it is the nature of the mind to incessantly think, become attached, and wander. True suffering and misery arise when the mind is unfocused and restless. Therefore, outward transformations—whether institutional, social, economic, technological, or ecological—cannot lead to genuine change if they are products of the same old, disordered mind. Logical reasoning thus suggests that unless there is a fundamental transformation within human consciousness, no lasting change can manifest in the external world (Behera, 2023).

The discussions in the earlier sections reaffirm that our current social and economic structures are fundamentally built upon acquisitiveness and competition—the relentless desire for more. Consequently, the crises confronting society today—whether social, economic, or ecological—can be traced back to a fundamental disorder within the human psyche. Various approaches, as discussed in Sect. 3, have been proposed over time to address this disorder, but they have largely fallen short of achieving the deep transformation required.

In recent years, however, new frameworks that integrate principles of social and ecological justice have begun to evolve. At their core lies the eco-spiritual worldview, which calls for a profound shift in human consciousness rooted in compassion, care, and interconnectedness. Ancient Indian philosophies offer rich insights into the understanding of the mind's workings, emphasizing that the quality of the external world is a direct reflection of the mind's state. In essence, without an inner revolution, outward reforms remain superficial.

As Fig. 3 illustrates, achieving true transformation in our external institutions, socio-economic structures, technologies, and ecological systems demands a prior transformation of the human psyche. External alterations alone, however significant, are still manifestations of the same conditioned mind. Therefore, meaningful and lasting change begins not with the world outside, but within the consciousness of each individual.

7.1 Transformation of the Human Mind

We propose that a profound transformation of the human mind (psyche) is essential for achieving fundamental change in the external world—across ecological, economic, and social systems. As Krishnamurti (2010b: 11–12) insightfully observed:

> All outward forms of change brought about by wars, revolutions, reformations, laws, and ideologies have failed completely to change the basic nature of man and therefore of society. As human beings living in this monstrously ugly world, let us ask ourselves, can this society, based on competition, brutality, and fear, come to an end? Not as an intellectual conception, not as a hope, but as an actual fact, so that the mind is made fresh, new, and innocent and can bring about a different world altogether? It can only happen, I think, if each one of us recognises the central fact that we, as individuals, as human beings, in whatever part of the world we happen to live or whatever culture we happen to belong to, are totally responsible for the whole state of the world.

It has become clear that for society to change, the transformation of the human mind must come first. Achieving a radical shift in human consciousness and emotion requires immense energy. To bring about such a transformation within oneself involves developing a deep sensitivity to everything around us. This means becoming acutely aware of every action we take. When individuals undergo such inward changes, they can contribute to creating a society that is profoundly compassionate and considerate—not only toward fellow human beings but also toward all forms of life on Earth. As Krishnamurti (1974) aptly stated, it is essential for everyone to cultivate compassion and love, as only these qualities can prevent humanity from descending into destruction and the degradation of the natural world:

> To have compassion means to have passion for all things, not just between two people, but for all human beings, for all things of the earth, the animals, the trees, everything the earth contains. When we have such compassion, we will not despoil the earth as we are doing now, and we will have no wars.

7.2 The Right Kind of Education

What can cultivate greater sensitivity and compassion in human beings? The right kind of education certainly has the potential to bring about a profound shift in both thinking and action. In this context, the key questions are: Does the current global education system adequately address how to live rightly? Does it foster a deeper understanding of the interconnectedness between human society and nature? Can the existing education system genuinely alter human behaviour? At present, education systems emphasize the accumulation of information, technological advancement, material success, and achievement, while fostering acquisitiveness and competition to secure employment. However, they often overlook the holistic development of individuals, the exploration of life's deeper meanings, and the importance of living in harmony with both society and the natural world. As Krishnamurti (1973: accessed online) poignantly stated:

> The function of education is to find out how to live differently, not merely to pass exams, to get a degree, become qualified in certain ways. It is to help you to face the world in a totally different, intelligent way, knowing you have to earn a livelihood, knowing all the responsibilities, the miseries of it all. Education is about understanding, with sensitivity and intelligence, of the whole of living - its extraordinary beauty, its sorrows, and joys.

Thus, the right kind of education is one that guides individuals in choosing a livelihood or vocation that fosters unity rather than division. It cultivates integrated action, intelligence, love, freedom, goodness, sensitivity, and the de-conditioning of the human mind. This education must also empower individuals to overcome fear, anxiety, anger, greed, violence, and sorrow. Once freed from these internal psychological disturbances, a person can navigate daily life with insight (the ability to observe without the interference of thought), love (a state of mind free from conflict and strife), and compassion (a deep care for all things on Earth). When individuals function from a place of insight, love, and compassion, they contribute to the creation of a caring, unified society and a more compassionate economic system.

7.3 *The Right Livelihood*

The world is primarily focused on securing a job or livelihood, often resulting in conflicts and suffering. It is crucial to understand what constitutes the right livelihood—one that does not generate conflict within society or harm the natural environment. Clearly, there are certain livelihoods and professions that perpetuate divisions and conflicts, both in society and with nature. As Krishnamurti (1947: 9) aptly stated:

> We are only too thankful to get what we can and blindly follow the economic system that is about us. But the questioner wants to know how to withdraw from exploitation and war. To withdraw from them he must not allow himself to be influenced, nor follow traditional occupation, nor must he be envious and ambitious. Many of us choose some profession because of tradition or because we are of a family of lawyers or soldiers or politicians or traders; or our greed for power and position dictates our occupation; ambition drives us to compete and be ruthless in our desire to succeed. So, he who would not exploit or contribute to the cause of war must cease to follow tradition, cease to be greedy, ambitious, self-seeking. If he abstains from these, he will naturally find right occupation. But though it is important and beneficial, right occupation is not an end in itself. You may have a right means of livelihood but if you are inwardly insufficient and poor you will be a source of misery to yourself and so to others; you will be thoughtless, violent, self-assertive. Without that inward freedom of Reality, you will have no joy, no peace. In the search and discovery

of that inward Reality alone can we be not only content with little, but aware of something that is beyond all measure. It is this which must be first sought out; then other things will come into being in its wake.

The Collected Works of J. Krishnamurti Volume IV (Page 9)

8 Concluding Remarks

This chapter offers a critical analysis of several economic development paradigms, identifying significant flaws within the prevailing economic system. It highlights the potential of alternative economic models, particularly those rooted in Gandhian and Buddhist economic and social principles. These models, grounded in values of equity, sustainability, and human dignity, offer promising pathways for promoting the collective wellbeing of both people and the planet. However, the argument is made, inspired by the philosophical writings of Jiddu Krishnamurti, that these models cannot be fully realized under the current global system due to the inward crisis of human consciousness. This crisis is a fundamental driver of the outward social, economic, and ecological crises we face today.

The key point of this argument is that the outward crises—such as social unrest, economic instability, and environmental degradation—are not isolated issues but are deeply connected to the psychological disorder within the human mind. Our current social and economic structures are built on a foundation of acquisitiveness, competition, and greed, all of which are fueled by inward psychological disorders. These disorders, which manifest as selfish desires and a constant pursuit of material wealth, create conflicts not only among human beings but also with the natural world. These inward conflicts contribute to the external strife that manifests in poverty, inequality, environmental destruction, and societal unrest.

To address these outward crises, a profound transformation of the human psyche is required. The systems that perpetuate these issues have been designed to satisfy narrow, individualistic interests. These systems are driven by the idea of maximizing self-interest, which may lead to short-term gains but ultimately results in long-term harm to society and the planet. For real change to occur, it is not enough to alter these outward structures alone. The transformation must begin at the individual level, with a radical shift in consciousness that moves away from selfishness, greed, and competition toward values of empathy, compassion, and social responsibility.

One of the most powerful ways to achieve this transformation is through education. The right kind of education can shape the human mind by helping individuals develop a deeper understanding of the significance of life, their interconnectedness with others, and their responsibility to the natural world. This education should not merely focus on acquiring information or technical skills; it must also foster emotional intelligence, ethical understanding, and a deep sense of compassion. By addressing the inward psychological disorders—such as fear, anxiety, anger, greed, and sorrow—education can help individuals develop the emotional resilience needed to navigate life in a way that is both fulfilling and harmonious with the world around them.

Furthermore, individuals whose minds are conditioned by a constant pursuit of pleasure through material consumption cannot form a meaningful and lasting relationship with nature. The unchecked desire for more—whether in terms of possessions, status, or power—leads to a misalignment of values. It results in the objectification of the environment and a distorted view of what is truly valuable. Nature, rather than being revered and protected, is exploited for its resources, with little regard for the long-term consequences of such actions. This disconnect from nature and the promotion of self-centered values contribute to the environmental and ecological crises that are threatening the survival of the planet.

Therefore, a crucial question arises: How can we de-condition the human mind to make it more sensitive, compassionate, and intelligent in its dealings with the world? This is the question that lies at the heart of the solutions to the global crises we face today. The answer to this question lies not only in cultural and traditional influences but also in the role of education and the immediate societal environment. The minds of individuals are shaped by the values, norms, and assumptions of the societies in which they live, and these values often encourage egotism, competition, and a narrow worldview that fails to recognize the interconnectedness of all life.

The current education system, in many parts of the world, tends to reinforce these limiting worldviews. It focuses heavily on acquiring knowledge, technological advancement, material success, and individual achievement. While these skills may be necessary for economic productivity, they often come at the expense of fostering deeper emotional and ethical intelligence. Education that promotes only the pursuit of material gain and societal status cannot equip individuals to navigate the complex challenges of the modern world. It fails to address the deeper questions of life, such as the nature of happiness, the interconnectedness of all beings, and the responsibility each individual has to society and the environment.

In contrast, an education that inspires individuals to care for one another and for the Earth offers the potential to de-condition and transform the human psyche. This type of education encourages individuals to develop a global perspective—one that transcends national, religious, political, and racial divisions—and helps them to cultivate values of empathy, compassion, and wisdom. By teaching individuals how to live harmoniously with others and with the natural world, such education can help bring about the radical transformation of society that is needed to address the ecological and economic crises we face.

Moreover, the right kind of education can help individuals make more ethical choices regarding their livelihoods. It can guide them in selecting professions and vocations that contribute positively to society and the environment, rather than perpetuating harmful practices that exacerbate the crises of inequality, environmental degradation, and social division. The ecological and economic challenges we face are not insurmountable, but they require a fundamental shift in how we think about success, wealth, and the purpose of human life.

The transformation of the human psyche, through the right kind of education, holds the potential to heal the deep divisions in society and restore harmony between humanity and the Earth. It is only when individuals are free from psychological bondage—such as fear, greed, and attachment—that they can function in their daily lives with insight, love, and compassion. When individuals act from a place of genuine understanding and empathy, they create a society and economy built on solidarity, care, and compassion. This is the kind of transformation that is urgently needed to address the crises facing the world today.

References

Agrawal, S., Dhayal, K. S., & Esposito, L. (2024). *From economic wealth to wellbeing: Exploring the importance of happiness economy for sustainable development through systematic literature review*. https://doi.org/10.1007/s11135-024-01892-z

Akolkar, A. H., Bharadwaj, K., Kulkarni, R., & Prasad, P. V. (2024). A comparative study of Kautilya Arthashastra and current Arthashastra. *International Journal for Multidisciplinary Research, 6*(5). https://doi.org/10.36948/ijfmr.2024.v06i05.28050

Allison, M. (2023). Socialism. *Victorian Literature and Culture, 51*(3), 505–508. https://doi.org/10.1017/S1060150323000517

Amore, R. C. (2024). *Buddhist economic ethics* (pp. 58–87). Oxford University Press. https://doi.org/10.1093/oxfordhb/9780192894328.013.3

Balasubramanian, S. (2022). *Kautilyanomics: For modern times*. Bloomsbury Publishing India Pvt.

Balasubramanian, S. (2024). *Dharmanomics: An indigenous and sustainable economic model*. Bloomsbury Publishing India Pvt.

Bawono, B., Surya, J., Vidya, D., & Suryanto, E. (2023). *Ekonomi dalam perspektif agama buddha*. https://doi.org/10.53565/patisambhida.v4i2.1057

Behera, B. (2012). *Consumption behaviour, economic growth, and sustainable well-being*. Paper presented in international conference "Planet Under Pressure: New Knowledge Towards Solution", held in London, United Kingdom, 26–29 March. https://doi.org/10.1007/978-981-99-2637-4_8

Behera, B. (2017). *Can sustainability problem be addressed without radically transforming the human psyche?* Paper presented in a seminar organised by the Institute of Geography and Sustainability (IGD), University of Laussane, Switzerland, 12 June (13.30–14.30) at Geopolis 3799. https://agenda.unil.ch/display?id=1496411660716.

Behera, B. (2023). A new economy for happiness. In S. Chetri, T. Dutta, M. K. Mandal, P. Patnaik, (Eds.), *Handbook of happiness*. Springer. https://doi.org/10.1007/978-981-99-2637-4_8

Bhat, R. M. (2023). Economic governance and public finance in kautilya's arthashastra. *International Journal of Social Science Educational Economics Agriculture Research and Technology (IJSET), 2*(7), 211–218. https://doi.org/10.54443/ijset.v2i7.175

Boyle, A. (2022). Buddhist economics: The global view. *Ethical Economy*, 339–360. https://doi.org/10.1007/978-3-031-10204-2_18

Brown, C. (2023). *Buddhist economics: The foundation for an equitable, sustainable, caring economy*. https://doi.org/10.15239/ycjcb.01.01.07

Burnham, T. C., & Phelan, J. (2023). Biological welfare economics: A natural science critique of normative economics. *Journal of Bioeconomics, 25*(1), 1–33.

Cardoso, E. L., Filho, H. R. F., Pinto, V. C. S., Souza, F. M. de, Ferreira, A. da O., Aguiar, E. S. de, Reis, A. C. M., & Luz, D. B. da. (2024). *Solidarity economic enterprise formed by waste pickers: A case study of the Águas Lindas waste pickers association, Belém, Pará*. https://doi.org/10.7769/gesec.v15i3.3650

Carnero, M. Á., Martínez, B., & Sánchez-Mangas, R. (2024). Time banking and social capital creation: A transaction data analysis. *Revista Galega de Economía*. https://doi.org/10.15304/rge.33.1.9263

Chetri, S. (2023). *Gross national happiness: The interdependent domains of happiness* (pp. 129–171). https://doi.org/10.1007/978-981-99-2637-4_7

Choudhury, M. A. (2020). The Meso-Economics of the Wellbeing Criterion. In *Islamic Economics as Mesoscience: A New Paradigm of Knowledge* (pp. 115–138). Singapore: Springer Singapore.

Das, S. K., & Mahapatra, S. D. (2012). *Business ethics and society: An insight to arthasastra.* https://indianresearchjournals.com/pdf/ijssir/2012/april/2.pdf

Dash, J. K. (2024). Gross national happiness (GNH): A way from national happiness to global happiness. *EPRA International Journal of Multidisciplinary Research.* https://doi.org/10.36713/epra16398

Diniz, E. H. (2023). *The journey from community bank to solidarity fintech through digital social currency* (pp. 35–42). Routledge eBooks. https://doi.org/10.4324/9781003317333-9

Dubois, L. (2021). The impact of solidarity economy on poverty: The case of public centres of solidarity economy in Bahia, Brazil. *World Development Perspectives, 23*, Article 100343.

Egorov, V. G., & Inshakov, A. A. (2023). The role of the solidarity economy in the development of society (on the example of Asian countries). *SHS Web of Conferences, 164*, 00045. https://doi.org/10.1051/shsconf/202316400045)

Egorov, V. G., Inshakov, A. A., Nogmova, A., & Alexandrova, E. V. (2023) Legal basis solidarity economy. *Постсоветский Материк, 3*, 51–78. https://doi.org/10.48137/23116412_2023_3_51

Etxeberria-Otadui, A. (2024). *Mondragon, Humanity at Work. 1*(3), 207–211. https://doi.org/10.46425/cjed1401039827

Gao, N. (2022). Buddhist economics: A guide to creating an equitable, sustainable, caring market economy. *Ethical Economy*, 361–377. https://doi.org/10.1007/978-3-031-10204-2_19

Gor, N. (2011). *The Dharma of capitalism: A guide to mindful decision making in the business of life.* Kogan Page Publishers.

Gordon-Nembhard, J., & Ifateyo, A. N. (2023). *African American solidarity economics and distributive justice* (pp. 105–112). Edward Elgar Publishing eBooks. https://doi.org/10.4337/9781803920924.00027

Gupta, R. K. (2024). Role of bharatiya gyan tradition in modern concept of arthashastra. *ShodhKosh Journal of Visual and Performing Arts, 5*(2). https://doi.org/10.29121/shodhkosh.v5.i2.2024.3232

Hawley, J. (1995). Dharmic management: A concept-based paper on inner truth at work. *Journal of Human Values, 1*(2), 239–248. https://doi.org/10.1177/097168589500100207

Hayden, A., & Dasilva, C. (2022). The wellbeing economy: Possibilities and limits in bringing sufficiency from the margins into the mainstream. *Frontiers in Sustainability, 3.* https://doi.org/10.3389/frsus.2022.966876

Hensher, M. (2024). Wellbeing economics—From slogan to discipline? Comment on "Can a Wellbeing Economy Save Us?". *International Journal of Health Policy and Management.* https://doi.org/10.34172/ijhpm.8871

Hough-Stewart, L., & Janoo, A. (2024). *Policy design for a wellbeing economy—Lessons from four city pilots.* https://doi.org/10.3138/jccpe-2023-0008

Hussen, M. A. (2004). *Principles of environmental economics.* Routledge.

Jakhar, B. R., Kumar, V., & Sharma, S. (2023). *Sanatan economics: An eternal economic order. 16*, 708–714. https://doi.org/10.2478/ers-2023-0043

Jena, L. K., & Behera, B. (2017). Environmental crisis and human wellbeing: A review. *International Journal of Development and Sustainability, 6*(8), 561–574.

Krishnamurti, J. (1945). *The collected works of J. Krishnamurti. The observer is the observed* (Vol. 4, p. 9). Krishnamurti Foundation America.

Krishnamurti, J. (1947). *The collected works of J. Krishnamurti. The observer is the observed* (Vol. 4, pp. 105). Krishnamurti Foundation America.

Krishnamurti, J. (1964). *The collected works of J. Krishnamurti. The dignity of living.* (Vol. 15, pp. 16). Krishnamurti Foundation America.

Krishnamurti, J. (1973). *The beginnings of learning (Part I)*: Chapter 13, School Dialogue, Brockwood Park, 17 June. https://jkrishnamurti.org/content/part-i-chapter-13-school-dialogue-brockwood-park-17th-june-1973.

Krishnamurti, J (1974). *Talks with American students*. Krishnamurti Foundation India, Chennai.
Krishnamurti, J. (1975). *Krishnamurti Journal*. Krishnamurti Foundation India. Chennai.
Krishnamurti, J. (2010a). *Freedom from the known*. Krishnamurti Foundation India, Chennai.
Krishnamurti, J. (2010b). *The first and last freedom*. Krishnamurti Foundation India, Chennai.
Krishnamurti, J. (2012). *Education and significance of life*. Krishnamurti Foundation India, Chennai.
Kumarappa, J. C. (1951). *Gandhian economic thought*. Sarva Seva Sangh Prakashan.
Kyyrönen, O. (2024). Wellbeing economy: Integrating sustainability and social welfare. *European Journal of Public Health, 34*(Supplement_3). https://doi.org/10.1093/eurpub/ckae144.625
Lane, T. (2017). How does happiness relate to economic behaviour? A review of the literature. *Journal of Behavioral and Experimental Economics, 68*, 62–78.
Lavie, D. (2022). Some imperfections of the modern economic system. Academy of Management Proceedings, p. 2.
Lazányi, O. (2022). *An Ecological Economics Inquiry into the Social and Solidarity Economy Insights from an Action Research*. Ph.D. Dissertation, Corvinus University of Budapest. https://doi.org/10.14267/phd.2022052
Legge, D. (2025). Navigating towards a wellbeing economy: Need for a robust theory of change; comment on "Can a Wellbeing Economy Save Us?". *International Journal of Health Policy and Management*. https://doi.org/10.34172/ijhpm.8873
Lemaître, A., & Helmsing, A. H. J. (2012). Solidarity economy in Brazil: Movement, discourse and practice analysis through a Polanyian understanding of the economy. *Journal of International Development, 24*(6), 745–762.
Leschke, M. (2016). A Critique of Welfare Economics. In *Order ethics: An ethical framework for the social market economy* (pp. 57–66). Cham: Springer International Publishing.
Maharsi, G., Badry, A. I., & Sadadi, P. (2023). Time banking as an alternative social movement: A case study of time banks in the United Kingdom. *International Journal of Social Service and Research*. https://doi.org/10.46799/ijssr.v3i8.487
Malov, A. N., & Popov, D. N. (2024). economic philosophy of an ancient India. *Гуманитарные Науки, 14*(4), 65–73. https://doi.org/10.26794/2226-7867-2024-14-4-65-73
Manrai, R., & Goel, U. (2017). Sustainable economic governance: Learning from Kautilya's Arthashastra. *International Journal of Indian Culture and Business Management, 15*(2), 241–253. https://doi.org/10.1504/IJICBM.2017.10007234
Marale, S. M. (2012). Shifting role of ecology in solving global environmental problems: Selected practical tools. *Environment, Development, and Sustainability, 14*(6), 869–884.
McDonald, E., & Richmond, K. (2019). Measuring Scotland's economic performance. *Fraser of Allander Economic Commentary, 43*(1). https://strathprints.strath.ac.uk/67584/
Newman, M. (2005). *Socialism: A very short introduction*. https://www.amazon.com/Socialism-Short-Introduction-Michael-Newman/dp/0192804316
Ongsod, B., & Braila, J. (2022). The national happiness index of Bhutan and its enlightenment. *Journal of Asia Social Science, 9*(1), 1–6. https://doi.org/10.51600/jass.2022.9.1.1
Parnell, E. (2021). The concept of the 'social and solidarity economy' damages the prospect of genuine self-help enterprises achieving a pivotal role within the economy. *Journal of Co-operative Studies, 54*(3), 67-72. https://doi.org/10.61869/BMPM7413
Patil, S. (2024). Review of Kautilya's arthashastra. *ShodhKosh Journal of Visual and Performing Arts*. https://doi.org/10.29121/shodhkosh.v5.i6.2024.2057
Perman, R., Ma, Y., McGilvray, J., & Common, M. (2003). *Natural resource and environmental economics*. Pearson Education Limited.
Priya, C. K. K. (2024). Strategic management in the Mauryan Era: Insights on economic prosperity and governance from Arthashastra. *African Journal of Biomedical Research*, 11061–11072. https://doi.org/10.53555/ajbr.v27i4s.5801
Robinson, M. (2023). Wellbeing economy in Australia: Progress, challenges and opportunities. *Health Promotion Journal of Australia*. https://doi.org/10.1002/hpja.770
Rosser, J. B., Jr., & Rosser, M. V. (2018). *Comparative economics in a transforming world economy*. MIT Press.

Schumacher, E. F. (1973). *Small is beautiful*. London: Abacus.

Shukla, S. (2024). Examining social and economic dynamics: A comparative analysis of Varna and Ashrama systems in Ancient India and Adam Smith's division of labor. *Journal of Leadership, Accountability, and Ethics, 21*(4). https://doi.org/10.33423/jlae.v21i4.7416

Sihag, B. S. (2005). Kautilya on ethics and economics. *Humanomics, 21*(3), 1–28. https://doi.org/10.1108/EB018902

Singh, R., & Alexandrova, A. (2020). Happiness economics as technocracy. *Behavioural Public Policy, 4*(2), 236–244.

Tian, G., & Chen, X. (2022). Economic analysis of the market economy system. In: *China's reform: History, logic, and future*. Springer. https://doi.org/10.1007/978-981-19-5470-2_9

Vinod, H. D. (2012). The Oxford Handbook of Hindu economics and business. *Research Papers in Economics*. https://econpapers.repec.org/RePEc:oxp:obooks:9780199782048

Vinod, H., & Deodhar, S. (2024). *Hindu economic ethics* (pp. 5–41). Oxford University Press. https://doi.org/10.1093/oxfordhb/9780192894328.013.1

Wallace, J. (2019). *Scotland: Wellbeing as performance management* (pp. 45–71). Palgrave Pivot, Cham. https://doi.org/10.1007/978-3-030-02230-3_3

WECD. (1987). From one earth to one world: An overview. In *World commission on environment and development (WECD)*, Oxford University Press.

Chapter 5
Economic System Beyond the Premise of Self-interest

1 Introduction

This book attempts to unravel the multiple dimensions of the present economic systems and the process adopted by them to prosper and sustain themselves. The objective is to simplify the complex process and present them in a way that makes for a broader understanding. The preceding chapters clearly demonstrated that the present growth centric economic systems are no longer sustainable given their excessive dependence on natural systems or natural capital. While growth-centric economic systems work on the economic premise of self-interest, this chapter aims to provide a more balanced economic development approach that could work beyond self-interest and greed. It is not fair to discard the growth-centric approach entirely. At the same time, it needs to balance multiple aspects and adopt sustainable practices and governance. That is balancing economic needs (decent living), social needs (equitable and fair distribution) and environmental or ecological needs (safeguarding the natural capital for future generations). Based on the review of various approaches (post growth, ancient and philosophical) an attempt is made in this chapter to explore how a more balanced approach or economic system that looks beyond the premise of self-interest could be made possible. The idea is not to provide an alternative framework or approach. At the same time, important ailments of the present system are identified.

2 Limits to Growth

The capitalist system is not only the longest-surviving economic system but also the most effective in terms of improving living standards and average prosperity. On the contrary, socialist/communist systems are short-lived as they fail to achieve or deliver the stated objectives of creating equitable and hunger-free societies. Some of the

V. R. Reddy and B. Behera, *Understanding the Current Global Socio-economic and Ecological Crises*, https://doi.org/10.1007/978-3-031-98600-0_5

countries have adopted or transferred to mixed economic systems, i.e., combination of public and private systems. The performance of these systems is mixed depending on their political system, governance, institutions, and implementation processes. Overall, democratic governance coupled with capitalist systems has been performing well. Socialist governance with a mixed economic model showed some exceptional performance in China, though sustaining such performance is being questioned of late.

Irrespective of their track record in reducing poverty and improving prosperity, all the economic systems have focused on financial capital at the cost of human and/ or natural capital. While improvements in financial capital are measured in terms of GDP, the costs associated with human and natural capital are rarely considered. As a result, the economic performance of a country is measured in terms of growth in financial capital, i.e., per capita income. Capitalist economic system is flexible and adoptive to the socioeconomic dynamics. Technological and institutional innovations have been the main drivers of capitalist growth for more than a century. As propounded in Schumpeter's theory of innovation (creative destruction) technological innovations drive economic growth through new products, processes, and organizational structures that disrupt existing markets, leading to a continuous cycle of creation and destruction. Every cycle of innovation has a lifespan and stops generating super profits beyond its life and till a new innovation hits the market. Whether it is industrialisation of 1930s, agricultural technologies during 1960s, economic reforms and globalisation (institutional changes) in the1980s, digital technologies (computerisation) in the 1990s, or the ongoing artificial intelligence (AI) wave or robotics or electric mobility (EB/EV) revolution have helped to reinvent and re-invigorating the capitalist system and sustaining it. New inventions have always given higher profit margins for their inventors and adopters and fuelled economic growth. This growth has been achieved at the cost of other resources or capitals i.e., human or natural. More importantly, excessive or high growth was achieved due to the imbalances and unsustainable use of natural resources.

While these imbalances are attributed to the selfish motives and profit maximisation tendencies of human behaviour. These economic traits have led to limitless growth that, in turn, fuelled 'human greed' overriding the 'human need.' This has led to increasing inequalities within and among the countries. Degradation of natural resources associated with high growth trajectories has further aggravated the inequalities, as vulnerable sections experience declining access to these resources (forests and other common pool resources). Technological innovations have hastened the degradation process. Even digital technologies like computers, mobile phones, etc., impose severe pressure on natural ecosystems (water, energy, etc.). Serious concerns are raised about the so-called environmentally friendly technologies like solar energy and electric car batteries. These technologies are likely to end up harming the natural systems with adverse impacts associated with lithium mining as well as silicon (solar panels) and lithium production and disposal (Boyden et al., 2016; Tawalbeh et al., 2020).

Despite the fact that environmental economics and environmental costs are well recognised and integrated into the mainstream policies of most developed countries, they are often used for arm twisting of the developing countries in the name of environmental protection. In the absence of a life-cycle assessment of these technologies and production processes, only the positive aspects are highlighted to the neglect of negative impacts. The negative environmental impacts of some of the earlier technologies (including the green revolution) are being experienced in most of the countries, especially in developing countries. In fact, developed countries like the USA are not honouring the climate targets. It may be noted that all these technologies are not entirely bad for the environment. However, the greed of the individuals and the economic system does not allow for a balanced and sustainable use of these technologies. Thus, it is not the problem of the technology per se. Rather it is the way these technologies are being used. Often some of these technologies are shifted to the developing countries in the name of development aid or to ensure cheap supplies to the developed countries. In order to facilitate such exchange global institutional and governance systems like free trade agreements, globalisation, etc., were evolved.

3 Changing World Economic Order

Along with technological innovations, the capitalist economic system gets stimulus from institutional interventions at the global level from time to time. These stimuli are in the form of free trade agreements, economic reforms, and policies that support the opening of markets. Globalization is the overarching market reform that allows the free flow of capital, labour, technology, goods, and services. Often these interventions are biased and used to the advantage of stronger and developed market economies. In the case of the free flow of technology and capital, developed countries benefited from the import of cheap goods and services, and developing countries benefited from employment generation. Unfortunately, even wars between countries and internal political disturbances are frequently used to revive the sagging capitalist markets and economies through export of defence equipment, ammunition, etc. Free flow of labour helps sustain their production and service sectors while helping the developing countries with capital flows (remittances). In the process, developed countries exported environmental degradation inflicting net damages on the developing countries. Even among the developed countries stronger economies benefited more (USA) when compared to weaker economies (European countries).

In order to overcome these anomalies, economic cartels or zones were formed viz., European union (EC) and BRICS (Brazil, Russia, India, China and South Africa). European developed economies have come together under EU while emerging economies have formed under BRICS. While EU formation has not proved to be very effective in addressing the socioeconomic problems of its member countries equally, as the stronger economies within the EU (Germany and France) seem to have taken advantage of the cartel. Such anomalies are yet to surface in the case of BRICS due to its recent nature. But, the recent developments clearly indicate that these cartels are hurting the economic interests of the USA. The USA is trying to deal with these cartels through reciprocal or even higher tariffs. Further, the USA is imposing restrictions on labour movements and immigration. Number of developed countries have introduced restrictive policies on labour movements, though some countries like Germany are freeing up to attract skilled workers from outside to meet the internal demand. On the whole, there are clear indications that free trade environment is being reversed globally. This could be the end of globalisation, which was introduced to booster the world economies. These developments are likely to further dampen the growth prospects.

The policy and institutional stimuli though help in sustaining the growth, they are temporary in nature and also dependent on the socioeconomic and political dynamics of different countries and regions. High growth trajectories are sustained for longer periods only through sustained degradation of natural resources. Decoupling of growth from resource degradation has not been established in any spacio-temporal context. Though technological innovations have proved to boost growth in the short run, most technologies are not environmentally friendly in the long run. And the recent innovations are no exceptions in this regard. The market economic system has been reinventing the growth process by placing a value on resource degradation. That is integration of environmental or resource degradation costs into the national income accounting process, viz., green accounting. Despite the availability of integrated project evaluation methods like life-cycle assessment, circular economy, etc., which incorporate environmental externalities in to costs and benefit flows, their adoption in the evaluation of real projects is very limited. Such an integration not only moderates growth but also affects profit margins adversely. This goes against the capitalist ethos of profit maximisation. As a result, green accounting or environmental cost–benefit assessments have remained at the academic level only rather than becoming a part of the policy discourse. Thus, the nature of growth would continue through the path of resource degradation. And the degradation trajectory is turning steeper and steeper over the years in order to maintain the growth rates.

4 Nature Based Solutions (NBS)

As the trade-offs between growth and resource degradation are not being addressed, some environmentally friendly approaches protect the ecosystem services through some direct approaches. Nature-based solutions (NBS) is one such method that directly emphasises the importance of protecting and strengthening natural capital. The term NBS emerged during the late 2000s', though it has its origins in the ecosystem services, natural capital-based approaches and agro-ecology. NBS is used as an umbrella concept encompassing all these aspects. NBS addresses major societal challenges like food and water security, climate impacts along with socio-economic development. Though NBS is being adopted in the context of climate risks it has a direct linkage with resource degradation and ecosystem services. NBS is based on the premise that healthy natural ecosystems produce a diverse range of services on which human wellbeing depends, viz., storing carbon, controlling floods and stabilizing shorelines and slopes to providing clean air and water, food, fuel, medicine and genetic resources. It is an 'overarching concept' encompassing other established 'nature-based' approaches like ecosystem-based adaptation/mitigation, eco-disaster risk reduction and green infrastructure, etc. (Seddon et al., 2020). In the context of agriculture NBS is defined as: "*the use of natural processes or elements to improve ecosystem functions of environments and landscapes affected by agricultural practices, and to enhance livelihoods and other social and cultural functions, over various temporal and spatial scales.*" (Cohen et al., 2016, p. 5). Agriculture directly provides number of ecosystem services, though the benefits of which are not valued. Farming communities benefit from these services directly in the form of commodity benefits (food) and they don't derive any benefit (value) from the non-commodity outputs (flood protection, water recharge, water purification, etc.), despite their contribution to their protection and provision.

Healthy soil, quality water, forests and bio-diversity are fundamental to agriculture as well as other production systems. Most countries face severe soil, water, biodiversity, forest and mineral resource degradation. The need to protect this natural capital and value their services is now recognised as critical for achieving sustainable development. NBS helps maximize the ability of natural capital base to enhance ecosystem services along with ensuring food production, climate adaptation, etc. (Matthews et al., 2019). NBS can be implemented from the local level to landscape level (watershed, river basin, etc.) and requires involvement of individuals to corporations and public agencies. Though NBS can be implemented at the individual farm/firm level, adopting them at scale maximizes their benefits (Cohen et al., 2016). Its implementation can be site-specific, incorporating local knowledge and culture, and integrated with other policies and interventions. The most important aspect of NBS is that it recognizes the trade-offs between present economic benefits and future production options and ecosystem services (Cohen et al., 2016). While the explicit focus of agroecology is soil health and crop systems (Ewert et al., 2023), NBS is much broader and considers other critical resources like water, ecosystem services, forests, minerals, bio-diversity, etc., and their associated externalities. NBS is proven to be cost-effective and help improve system resilience. NBS interventions are amenable to policy regulations and economic incentives/disincentives (Iseman & Miralles-Wilhelm, 2021).

Economic policies need to move towards NBS that focus on strengthening the natural capital base, resource conservation and governance strategies. NBS requires and calls for 'Comprehensive and Integrated Research'. Natural resources and environmental research need to adopt inter and trans-disciplinary approaches rather than working in silos. Integration is required at the resource level, viz., land, water, forests, minerals, etc., which demands integration of disciplines (bio-physical sciences—soils, hydrogeology, climatology, social sciences, behavioral sciences, etc.). NBS provides a framework for integrating the commodity and non-commodity goods and services of production systems. This would help in arriving at the real contribution of each sector. While the process is complex, minimum understanding of all the related aspects is necessary for designing and implementing informed and effective policies.

NBS is part of sustainable practices, viz., green infrastructure, green energy, soil nutrient management, conservation agriculture, etc., (Iseman & Miralles-Wilhelm,

2021). The objective of sustainable development is to bring a balance between natural and financial capital. Increased green initiatives and production processes can reduce ecological footprint. The activities include increased renewable energy production, increased forest and soil organic cover, increased crop and species diversification (diverse cropping, flora and fauna species). Though these practices are expected to increase soil organic matter, forest cover, bio-diversity and sediment retention, these principles are often not adopted comprehensively. Despite their promotion for about two decades, their adaptation rates are quite low due to long gestation periods. Systematic promotional activities may help their adoption, as the practices can reduce operational costs, increase resource use efficiency, natural system stability, etc., (Corsi & Muminjanov, 2019). Despite the significant direct and indirect benefits, from individual producers to the national level, associated with NBS, the adaptation of these practices is neither easy nor smooth. This is mainly due to the time lag between adaptation and benefit flows. More importantly, NBS is likely to increase costs and reduce the profit margins in the short run. This is a major constraint for adapting to NBS and needs policy support in the form of incentives for adopting NBS.

Profitability is the key to production decisions which are often high on returns and short-term in nature. Placing value and paying for their multifunctional/non-commodity outputs could help them move towards long-term NBS benefits. For, NBS needs to make an economic argument that goes beyond social or public benefits (ecosystem services) for adoption by producers and profitability is essential for incentivizing adaptation of practices that have long gestation (Huntley et al., 2020). NBS in principle is capable of addressing the trade-offs and putting economic values on non-commodity outputs. Even in the case of profitable NBS options, the number of non-economic factors like awareness, lack of skills, convenience, etc., hinder adoption levels.

5 Beyond the Premise of Self-interest

All the measures or approaches (technologies, institutions, NBS) are temporary in nature with degree difference in terms of their longevity of impacts. As discussed in the preceding chapters though some of the post/de-growth models have been advocating changes in the consumption patterns, quantity as well as composition, for sustaining the earth's present as well as future generations. These models expect that these changes could be achieved through cosmetic policy changes. However, such changes are fundamental and strongly linked to individual preferences requiring behavioral changes. These behavioral changes are required at production, consumption and governance (policy) levels in order to make the economic system effective and sustainable. In what fallows some of the important and required behavioral changes are discussed. Most of these behavioral patterns are drawn from ancient as well as recent Indian philosophies.

5.1 Need Versus Greed

Excessive accumulation and the exploitative nature of the present growth models are the main reasons for their poor performance and sustenance. The growth centric models have failed in terms of reducing extreme global poverty. These approaches have fuelled greed in human minds by-passing the importance of meeting basic needs. This has resulted in excessive exploitation of natural systems beyond sustainable limits. This is omnipresent across the sectors (agriculture and manufacturing) and among consumers and producers. While the exploitation of natural capital is the basis for the modern growth trajectory, its fast degradation is adversely impacting the growth potential. The growth is likely to reach its limits very soon, as already evident from experiences of some regions and sectors. Ignoring the Gandhian principle 'nature has enough to meet the humanities needs but not the greed' is the main reason for the present state of environment and human well-being.

Moderating the greed in human behaviour is fundamental to achieving the true sustainable development in its word and spirit. This calls for a balanced approach to production and consumption by keeping the natural resource use at sustainable levels. This would slow down the growth rate, especially among the developing countries. At the same time the balanced approach is likely to boost green growth i.e., adopting green accounting of national income. In the process developing counties may have to delay their aspirations to achieve the status of developed nations. For instance, India is aiming to achieve the status of developed economy requiring high GDP growth rates (above 10%) for longer periods. Environmental costs associated with such high growth rates are conveniently ignored. The pressure on the natural systems (soils, forests, and groundwater) and its impact on agriculture is very much evident in a number of regions in India. Declining farm viability and performance of agriculture not only impacts the well-being of the majority population but also impacts the overall growth rate. More importantly, it is likely to have an adverse impact on the food security of the country. Thus, adopting sustainable farm management practices with moderate growth rather than greedy extractive practices like excessive chemical use that would make the soils and water redundant and poisonous is a prudent economic decision.

5.2 Superprofits and Inequalities

Human greed at the production level has translated into super profits due to the under-valuation of natural resources and other factors of production. Besides, in the absence of environmental regulations in developing countries, the adoption of environment-friendly technologies is marginal. This enhances the profit margins of the producers, including the multinationals. Given that these super profits are mostly garnered by the major corporates with monopoly market power coupled with unfair distribution of shares to other factors of production has resulted in widening income inequalities across the socioeconomic ladder over the years. Corporate margins have been growing at a much faster rate than other sections of the population in recent years. This observation holds good across developed and developing countries. Of late, corporations are becoming powerful enough to dictate political entrepreneurs and policy makers so that they can make more profits (crony capitalism).

On the contrary, inflationary trends, especially food, during the post COVID 19 and the increased geopolitical disturbances have further aggravated the inequalities. In fact, food inflation and food insecurity in number of countries is attributed to the resource degradation and policy neglect of the less profitable farm sector. For, these developments have helped increase the profit margins of the corporates and other producers and reduced the disposable incomes of the wage earners and informal producers (agriculture). In most countries health and other service costs have increased substantially in recent years. This has resulted in the slipping down of the middle and upper middle classes to the lower status i.e., lower middle and middle class respectively. That is only ten percent of the total population is enjoying the fruits of so-called high growth rates while the remaining 90% are increasingly deprived of basic services. That is trickle down effects have been marginal. It may be noted that extreme poverty and destitution are not conspicuous but decent living standards are deteriorating even in some of the developed countries.

Such a sharp rise in inequalities may not be sustained in the long run which was amply demonstrated in economic history, especially in the case of feudalism, the French Revolution, etc. Time is approaching for the neo-liberalism too, unless appropriate and effective corrections are made to the economic system. While neoliberalism emphasises the role of markets in bringing behavioural changes, market systems have often designed (policy interventions) to favour the producers rather than consumers. Global experience shows that perfect competition is either utopian or short-lived. There is enough evidence to suggest that market mechanisms are incapable of correcting natural resource degradation anomalies (including externalities), especially in the case of a nexus between corporations and policy makers.

The contradictions of the growth-centric economic system could only be overcome through fundamental behavioural changes at the producer as well as policy levels. The self-interest-driven profit maximisation behaviours need to be moderated and replaced with more benevolent approach of sharing and inclusiveness of all factors of production. Though these changes could come across only through internal metamorphosis, policies and governance institutions could help in bringing the changes to some extent. Policy environment should pave the way for transparency and accountability about what is produced? and how it is produced? Policies can prioritise what to produce and guide the production processes so that maladaptation of new technologies could be minimised. Policies should help moderate excessive profits through dis-incentive mechanisms. Governance institutions should be capable of enforcing the policy framework.

5.3 External Versus Internal Happiness

Neo-liberal economic systems have driven individuals towards emphasising the importance of material goods and commodification. Consumerism is the main driver of capitalist growth. Materialist or external happiness has been embedding in the human psyche mainly due to the external and surrounding environment. Internal happiness comes through inner self-development through the accumulation of knowledge rather than the accumulation of material goods. Internal happiness has its origins in Hindu and Buddhist philosophies that emphasise self-enlightenment. Self-contentment is integral to internal happiness. In the absence of internal happiness and contentment, economic systems are caught in a vicious circle of material goods, commodification and consumerism. Greed often takes over self-contentment in an economic environment that is driven by self-interest and materialistic nature.

Though internal happiness sounds philosophical in more ways and appears impractical, there are practical examples of adopting these principles in different countries like Bhutan, Finland, Norway and Denmark. While it is understandable that developed Scandinavian countries could adopt such systems due to their economic status and affordability. These countries focus on sustainable resource management practices. For, these countries have obtained a level of well-being and hence could slow down on their growth trajectories without sacrificing much on their well-being or living standards. However, a poor South Asian country like Bhutan adopting and promoting a sustainable economic system is rooted in its cultural ethos that originates from Buddhist philosophies. And how these ethos and philosophies are being carried forward needs further probing of their system, specifically education system.

Despite being a small country, Bhutan is striving to balance its natural resources and food security. Its policy focus is on sustainable development through the protection of its natural capital heritage, despite the fact that it is neither self-sufficient nor food secure. The primary sector in Bhutan is termed as the renewable natural resources (RNR) sector, which includes agriculture, livestock, forestry, and mining sub-sectors. Bhutan is the only country in South Asia that maintains 70% of its geographical area under forests despite the constitutional requirement of 60%. Bhutan has formulated the number of policies in order to achieve the overall objectives of food security and sustaining the natural resource base. The economic development policy of Bhutan envisions a green and self-reliant economy. Materialistic tendencies and consumerism are conspicuous by their absence. Even the visitors (tourists) can experience the contrast between Bhutan and its neighbouring countries on their first outing.

However, the situation appears to be changing of late i.e., increasing preference towards external happiness at the individual level. This is mainly attributed to the exposure to the outside world, especially to the glitzes of neo-liberal consumerism and material goods. For, neo-liberal economic systems can spread their tentacles wide and deep. It is very difficult to expect the survival of small islands of sustainable practices, especially in poor countries like Bhutan. It would be meaningful to explore the potential and possibilities for sustaining such systems striving for internal happiness.

5.4 Right Education and Livelihoods

Apart from cultural practices and religious beliefs education shapes human behaviour. Often education systems and curriculum are designed reflecting the country or region-specific cultural practices and religious beliefs. These systems sustain themselves as long as they are isolated from external influences. For instance, the Indian education system was totally transformed from its ancient values and practices after the British takeover. By the end of the 200 years of British rule, India more or less lost its original education system. This is the case with most of the developing countries that came under imperialist influences. Over the years education systems across the world have become mono-cultural, i.e., western education system. Self-interest and neo-liberal thinking are integral to the Western education system. Such an education has obliterated the ancient wisdom from the resident countries. This is more conspicuous in the case of countries with rich ancient histories and cultural practices, viz., India/South Asia, etc.

Thus, the right education imbibes the right values in individuals and society. This is not to say that the Western education system is not right or bad. Western system is more of individual centric and emphasises that individual growth leads to overall growth and well-being. On the other hand, ancient education was community-centric, and people were taught to focus on the common good. Inclusivity is central to the ancient education system as equal importance is given to natural resources, livestock as well as wildlife. The inclusivity not only ensures due share to each and every factor of production but also keeps the resource trade-offs at moderate levels. While inequalities are accepted as a necessary evil, widening/deepening inequalities are not socially encouraged. Right education also paves the way for designing sustainable policies and creating appropriate policy environment that ensures sustainable development.

The present education system is viewed as an investment proposition that has its origins in the theory of human capital, another invention of neo-classical economics. Education is no longer sought as a tool for acquiring knowledge or internal happiness. The preference for specific specialisation is determined by its potential for market returns rather than student's aptitude or interest. The present system ensures higher incomes but not the right kind of livelihoods that are pursued with passion by individuals. The education system can make or break the social fabric of a nation. Designing and adopting an appropriate education system that considers the importance of conserving the natural systems and imbibes inclusive and sharing nature in human mindset is a necessary condition for attaining sustainable development. Sustainable development needs fundamental changes in human behaviour and only right education can help transform the human mind and behaviour.

As Jiddu Krishnamurti emphasized education as a means to free the mind from conditioning, fostering independent inquiry rather than conformity. Hence, it is essential to integrate this into modern economic systems by developing educational institutions that can inculcate self-awareness, creativity, and critical thinking among students instead of competition-driven success and encourage students to question societal norms, consumerism, and unsustainable economic practices. It is also paramount to teach the principles of Dharmic capitalism—where wealth creation aligns with responsibility, service, and sustainability. Having mindfulness in economic decision-making is key to fostering a sustainable future. Hence, it is essential to incorporate practices like meditation and self-inquiry to help future leaders make decisions based on awareness rather than greed.

The pitfalls of the current market economic systems identified in this book can be addressed by the inclusion of Dharmic capitalism into the system because Dharmic capitalism aligns business and economic activities with higher ethical and spiritual values. Dharmic capitalism tends to shift from profit-maximization to purpose-driven business, which encourages companies to see wealth creation as a means of *service* rather than an end in itself. More importantly, it stresses self-awareness in leadership where business leaders should practice self-inquiry, understanding their motivations and impact on society. Embedding long-term environmental, social, and cultural sustainability into corporate strategies is another important aspect of Dharmic capitalism. For Dharmic capitalists, wealth is a means for societal well-being, not for individual gratifications, hence encouraging reinvestment into community welfare, education, and well-being rather than excessive accumulation.

What is making the current economic system extremely weak and vulnerable, both socioeconomically and ecologically, is the blind pursuit of materialism. Therefore, reforms at all levels are urgently necessary. An urgent corporate governance reform is critical, and it should shift the focus from short-term profits to *long-term holistic well-being* (employees, environment, stakeholders). Fostering social enterprises & conscious businesses in the economies where business models are encouraged to balance profit with social responsibility. Alternative economic models such as, circular economy principles, cooperatives, and trust-based finance systems in which businesses prioritize trust over cutthroat competition should be promoted aggressively. Finally, steps should be taken to enhance consumer awareness in terms of educating consumers on mindful consumption, ethical purchasing, and their role in shaping market demand.

6 Conclusions

As the title suggests the objective of the book is not to suggest an alternative economic system or a framework that would address the problems associated with the current economic system. As the growth-centric approach has been pushing societies towards extreme inequalities and crony capitalism is gaining prominence. In the process, the free market economy, which is fundamental to the capitalist system, is increasingly becoming dysfunctional or ineffective in efficiently allocating resources. This would push resource degradation to unsustainable levels. Besides, the extreme inequalities would disturb the social fabric leading to socioeconomic disturbances. Such disturbances are already evident in the number of countries.

The present review identifies important weaknesses (greed, inequalities, and external happiness) of the present system that need to be addressed in order to restrict the present system from digressing from the self-destructive path. These undesirable traits of the human psyche are integral to the socioeconomic progress of any system, For, self-interest is the basis for economic progress over millennia. However, they need to be moderated and keep the balance among different traits of human behaviour. This can be achieved only through the right kind of education that emphasises cultural practices and philosophical ethos that are region-specific. The neo-liberal approach to education needs to be shifted away from materialism and external happiness and move towards internal happiness (accumulation of knowledge). Transformation of the human mind and behaviour is only possible through the right education.

References

Boyden, A., Soo, V. K., & Doolan, M. (2016). *The environmental impacts of recycling portable lithium-ion batteries*. http://creativecommons.org/licenses/by-nc-nd/4.0/

Cohen-Shacham, E., Walters, G., Janzen, C., & Maginnis, S. (2016). *Nature-based solutions to address global societal challenges*. https://portals.iucn.org/library/sites/library/files/documents/2016-036.pdf

Corsi, S., & Muminjanov, H. (2019). *Conservation agriculture: training guide for extension agents and farmers in Eastern Europe and Central Asia*. Available at http://www.fao.org/3/i7154en/i7154en.pd

Ewert, F., Baatz, R., & Finger, R. (2023). Agroecology for a sustainable agriculture and food system: From local solutions to large-scale adoption. *Annual Review of Resource Economics, 15*, 22.1–22.31. https://doi.org/10.1146/annurev-resource-102422-090105

Huntley Lafave, C., Ahren Renton, C., & Sierks, K. (2020). Show me the money: Making a downstream market for diverse regenerative crops. In: *Conservation finance network* [online]. https://www.conservationfinancenetwork.org/2020/04/16/show-me-the-money-making-downstream-market-for-diverseregenerative-crops

Iseman, T., & Miralles-Wilhelm, F. (2021). *Nature-based solutions in agriculture—The case and pathway for adoption*. Virginia. FAO and The Nature Conservancy. https://doi.org/10.4060/cb3141en

Matthews, J., Matthews, N., Simmons, E., & Vigerstol, K. (2019). *Wellspring: Source water resilience and climate adaptation*. https://www.nature.org/content/dam/tnc/nature/en/documents/Wellspring_FULL_Report_2019.pdf

Seddon, N., Chausson, A., Berry, P., Girardin, C. A. J., Smith, A., & Turner, B. (2020). Understanding the value and limits of nature-based solutions to climate change and other global challenges. *Philosophical Transactions of the Royal Society B*. https://doi.org/10.1098/rstb.2019.0120

Tawalbeh, M., Al-Othman, A., Kafiah, F., et al. (2020). Environmental impacts of solar photovoltaic systems: A critical review of recent progress and future outlook. *Science of the Total Environment*. https://doi.org/10.1016/j.scitotenv.2020.143528

The manufacturer's authorised representative in the EU is Springer Nature Customer Service Centre GmbH, Europaplatz 3, 69115 Heidelberg, Germany. If you have any concerns regarding our products, please contact ProductSafety@springernature.com

Printed and bound by CPI Group (UK) Ltd, Croydon, CR0 4YY
01/10/2026
02225977-0001